DEJA REVIEW™
Physiology

NOTICE

Medicine is an ever-changing science. As new research and clinical experience broaden our knowledge, changes in treatment and drug therapy are required. The authors and the publisher of this work have checked with sources believed to be reliable in their efforts to provide information that is complete and generally in accord with the standards accepted at the time of publication. However, in view of the possibility of human error or changes in medical sciences, neither the authors nor the publisher nor any other party who has been involved in the preparation or publication of this work warrants that the information contained herein is in every respect accurate or complete, and they disclaim all responsibility for any errors or omissions or for the results obtained from use of the information contained in this work. Readers are encouraged to confirm the information contained herein with other sources. For example and in particular, readers are advised to check the product information sheet included in the package of each drug they plan to administer to be certain that the information contained in this work is accurate and that changes have not been made in the recommended dose or in the contraindications for administration. This recommendation is of particular importance in connection with new or infrequently used drugs.

DEJA REVIEW™
Physiology

David W. Lin, MD, ME
Tufts University School of Medicine
Class of 2006

New York Chicago San Francisco Lisbon London Madrid Mexico City
Milan New Delhi San Juan Seoul Singapore Sydney Toronto

The **McGraw·Hill** Companies

Deja Review™: Physiology

Copyright © 2007 by The McGraw-Hill Companies, Inc. All rights reserved. Printed in the United States of America. Except as permitted under the United States Copyright Act of 1976, no part of this publication may be reproduced or distributed in any form or by any means, or stored in a database or retrieval system, without the prior written permission of the publisher.

Deja Review™ is a trademark of the McGraw-Hill Companies, Inc.

1 2 3 4 5 6 7 8 9 0   DOC/DOC   0 9 8 7 6

ISBN-10: 0-07-147510-9
ISBN-13: 978-0-07-147510-5

This book was set in Palatino by International Typesetting and Composition.
The editor was Patrick Carr.
The production supervisor was Catherine Saggese.
The text was designed by Marsha Cohen/Parallelogram.
Project management was provided by International Typesetting and Composition.
RR Donnelley was printer and binder.

This book is printed on acid-free paper.

INTERNATIONAL EDITION ISBN-10: 0-07-110491-7; ISBN-13: 978-0-07-110491-3

Copyright © 2007. Exclusive rights by The McGraw-Hill Companies, Inc., for manufacture and export. This book cannot be re-exported from the country to which it is consigned by McGraw-Hill. The International Edition is not available in North America.

Library of Congress Cataloging-in-Publication Data

Deja review : physiology / David W. Lin ...[et al.].
   p. ; cm. — (Deja review)
  ISBN 0-07-147510-9—ISBN 0-07-147495-1
   1. Human physiology—Examinations, questions, etc. 2. Physicians—Licenses—United States—Examinations—Study guides. I. Lin, David W. II. Title: Physiology. III. Series.
  [DNLM: 1. Physiology—Examination Questions. QT 18.2 D326 2006]
QP40.D45 2006
612.0076—dc22

2006045667

Thanks mom, dad, and Jen for always being there to love and support me. Thanks to all my friends for helping me keep things in perspective. And a special thanks to Xuan for her love, support, and patience.
—David W. Lin, MD, ME

Thank you to my family and friends for all of the years of love and support. And a special thanks to my wife, Miki, for the love to accomplish anything and my daughter, Madeleine, who reminds me everyday that life is wondrous.
—Spencer C. Behr, MD

I would like to thank my family for always being there for me. I would also like to thank Dave for his love and support.
—Xuan M. Luu, MD

I would like to thank my family and friends for all their love, support, and patience. I would also like to thank my fellow collaborators especially Dave Lin, for this great opportunity.
—Ekwutosi M. Okoroh, MD, MPH

To Al, for giving me unconditional support. To my parents, for teaching me hard work. To my friends, for making me laugh through medical school.
—Trevin C. Lau, MD

Contents

Contributing Authors	ix
Acknowledgments	x
Preface	xi

Chapter 1   **GENERAL PHYSIOLOGY**   1
Cell Membranes and Transport Across Membranes / 1
NA^+-K^+ Pump / 3
Intracellular Connections / 3
Osmosis / 4
Membrane Potentials / 4
Neuromuscular and Synaptic Transmission / 8
Skeletal and Smooth Muscle / 13

Chapter 2   **NEUROPHYSIOLOGY**   19
Autonomic Nervous System / 19
Sensory Systems / 22
Vision / 25
Hearing and Balance / 29
Taste and Olfaction / 32
Motor Systems / 35
Cerebral Cortex Functions / 40
Temperature Regulation / 42

Chapter 3   **CARDIOVASCULAR PHYSIOLOGY**   45
Vasculature / 45
Hemodynamics / 46
Electrophysiology—Electrocardiogram / 47
Electrophysiology—Cardiac Action Potential / 49
Electrophysiology—Pacemaker potential / 50
Electrophysiology—Conduction and Excitability / 52
Cardiac Muscle / 54
Contractility / 55
Starling Relationships / 56
Cardiac Output / 58
Cardiac Cycle / 59
Heart Sounds and Murmurs / 63
Arterial Pressure Regulation / 64
Special Circulations / 66
Microcirculation and Lymphatics / 67

Chapter 4   RESPIRATORY PHYSIOLOGY   71
Lung Volumes and Capacities / 71
Respiratory Mechanics / 74
Gas Exchange and Transport / 78
Pulmonary Circulation / 84
Ventilation and Perfusion / 86
Respiratory Control / 87
Response to Stress / 89

Chapter 5   RENAL AND ACID-BASE PHYSIOLOGY   93
Body Fluids / 93
Renal Filtration and Blood Flow / 94
Renal Secretion, Reabsorption, and Excretion / 97
Other Electrolytes / 105
Dilution and Concentration of Urine / 106
Acid-Base / 108

Chapter 6   GASTROINTESTINAL PHYSIOLOGY   117
Gastrointestinal Tract / 117
Gastrointestinal Hormones / 120
Gastrointestinal Motility / 124
Gastrointestinal Secretions / 129
Gastrointestinal Digestion / 134
Gastrointestinal Absorption / 136

Chapter 7   ENDOCRINE AND REPRODUCTIVE PHYSIOLOGY   141
Hormones / 141
Hypothalamus and Pituitary Gland / 145
Adrenal Gland / 150
Testis / 154
Ovary and Placenta / 158
Pancreas / 162
Thyroid Gland / 165
Parathyroid Gland / 168

Chapter 8   CLINICAL VIGNETTES   173

Suggested Readings   197
Index   199

Contributing Authors

Spencer C. Behr, MD
Resident Radiologist
Department of Radiology
Lahey Clinic
Burlington, MA
Class of 2005
Tufts University School of Medicine
Boston, MA

Trevin C. Lau, MD
Resident Obstetrician/Gynecologist
Department of Obstetrics and Gynecology
Brigham & Women's Hospital
Boston, MA
Class of 2006
Tufts University School of Medicine
Boston, MA

David W. Lin, MD, ME
Resident Surgeon
Department of Surgery
Baystate Medical Center
Springfield, MA
Class of 2006
Tufts University School of Medicine
Boston, MA

Xuan M. Luu, MD
Resident Surgeon
Department of Surgery
University of Connecticut Health Center
John Dempsey Hospital
Farmington, CT
Class of 2006
Tufts University School of Medicine
Boston, MA

Ekwutosi M. Okoroh, MD, MPH
Resident Obstetrician/Gynecologist
Department of Obstetrics and Gynecology
Maricopa Medical Center
Phoenix, AZ
Class of 2006
Tufts University School of Medicine
Boston, MA

Acknowledgments

The authors would like to recognize the faculty and staff at Tufts University School of Medicine for their endless commitment to education. Without them the knowledge base for this book would not have been developed-thanks for your teachings, support, and encouragement. We would also like to thank Dr. Charles R. Ashby, Jr., and his colleagues at St. John's University of Pharmacy for reviewing the factual content of our material, as well as the medical students who used this text in preparation for their boards and provided critical feedback. Finally, special thanks need to go to Vastavikta Sharma and her colleagues at ITC and especially our editor, Marsha Loeb, for her patience and guidance throughout this project.

Preface

The main objective of a medical student preparing for Step 1 of the United States Medical Licensing Examination (USMLE) is to commit a vast body of knowledge to memory. Having recently prepared for Step 1, we realize how daunting this task can be. We feel there are two main principles that will allow you to be successful in your preparations for Step 1: (1) repetition of key facts and (2) using review questions to gauge your comprehension and memory. The *Deja Review* series is a unique resource that has been designed to allow you to review the essential facts and determine your level of knowledge on the different subjects tested on Step 1. We also know, from experience, that building a solid foundation in the basic sciences (like with this physiology review book) will allow you to make a smooth transition into the clinical years of medical school.

ORGANIZATION

All concepts are presented in a question and answer format that covers the key facts on hundreds of commonly tested physiology topics that may appear on the USMLE Step 1 exam. The material is divided into chapters organized by physiologic systems along with a special chapter at the end that incorporates the material with their clinical presentation and relevance. Special emphasis has been placed on the pathways that govern the physiologic processes, as this area is vital for comprehension of how the various systems work, how they can go awry, and is the basis for the majority of questions in both your medical school course and Step 1.

This question and answer format has several important advantages:
- It provides a rapid, straightforward way for you to assess your strengths and weaknesses.
- It allows you to efficiently review and commit to memory a large body of information.
- It offers a break from tedious, convoluted multiple-choice questions.
- The clinical vignettes that are incorporated, exposes you to the prototypic presentation of diseases classically tested on the USMLE Step 1.
- It serves as a quick, last-minute review of high-yield facts.

The compact, condensed design of the book is conducive to studying on the go, especially during any downtime throughout your day.

HOW TO USE THIS BOOK

This text has been sampled by a number of medical students who found it to be an essential part of their preparation for the physiology questions on Step 1, in addition to their course examinations. Remember, this text is not intended to replace comprehensive textbooks, course packets, or lectures. It is simply intended to serve as a supplement to your studies during your medical physiology course and throughout your preparation for Step 1. We encourage you to begin using this book early in your first year to reinforce topics covered on your course examinations. We also recommend covering up the answers (instead of just reading both the question and the answer) and quizzing yourself or even your classmates. For a greater challenge, try covering up the questions!

However, you choose to study, we hope you find this resource helpful throughout your preclinical years and during your preparation for USLME Step 1. Best of Luck!

David W. Lin, MD
Spencer C. Behr, MD
Xuan M. Luu, MD
Trevin C. Lau, MD
Ekwutosi Okoroh, MD

CHAPTER 1

General Physiology

CELL MEMBRANES AND TRANSPORT ACROSS MEMBRANES

What are the components of a cell membrane?	1. Cholesterol 2. Phsopholipids 3. Sphingolipids 4. Glycolipids 5. Proteins
What is the role of cholesterol in the cell membrane?	Promotes membrane stability
What are the components of the phsopholipids in the cell membrane?	1. Glycerol backbone (hydrophilic) 2. Fatty acid chains (hydrophobic)
How are the phsopholipids in a cell membrane arranged?	Bilayer
What are the types of cell membrane proteins?	Integral and peripheral
How are integral and peripheral proteins different?	Integral: span entire membrane Peripheral: located on either side of the membrane
How are integral proteins attached to the membrane?	Hydrophobic interactions with the phospholipid bilayer
Give some examples of both types of proteins.	Integral: ion channels, transport proteins Peripheral: hormone receptors
How do lipid-soluble substances move across cell membranes?	Dissolving across the hydrophobic lipid bilayer via passive diffusion
How do water-soluble substances move across cell membranes?	Cross through water-filled channels or transported by carriers, as they cannot dissolve in the lipid bilayer
What are the different types of transport across a cell membrane?	1. Simple diffusion 2. Facilitated diffusion 3. Primary active transport 4. Cotransport 5. Counter-transport

Which types of transport are carrier mediated?	1. Facilitated diffusion 2. Primary active transport 3. Cotransport 4. Counter-transport
What types of transport require metabolic energy?	1. Primary active transport 2. Cotransport 3. Counter-transport
What is the direction of solute movement relative to the electrochemical gradient in the following types of transport:	
Simple diffusion	Downhill
Facilitated diffusion	Downhill
Primary active transport	Uphill
What is the equation to measure diffusion?	Fick's law: $$J = -PA(C_1 - C_2)$$ where J = flux (mmol/s), P = permeability (cm/s), A = area (cm^2), C_1 = concentration of substance 1 (mmol/L), C_2 = concentration of substance 2
What does permeability describe?	Ease with which a solute is able to diffuse through a membrane
What factors can increase permeability?	1. Increased oil/water partition coefficient 2. Decreased size of solute 3. Decreased membrane thickness
What are the characteristics that are important in carrier-mediated transport?	1. Stereospecificity 2. Saturation 3. Competition
What is the transport rate when the carriers are saturated?	Transport maximum (T_m)
Which is faster, diffusion or facilitated diffusion?	Facilitated diffusion, as it is carrier-mediated
What type of transport are cotransport and counter-transport?	Secondary active transport
How are cotransport and counter-transport mediated?	Transport of two or more solutes is coupled with one of the solutes transported down an electrochemical gradient to provide the energy for another up against a gradient

What is cotransport?	Movement of 2$^+$ solutes in the same direction across a cell membrane
What is counter-transport?	Movement of 2$^+$ solutes in the opposite direction across a cell membrane

NA$^+$-K$^+$ PUMP

What is the Na$^+$-K$^+$ pump also known as?	Na$^+$-K$^+$-ATPase
What is the function of the pump?	Extrudes 3 Na$^+$ from within the cell and takes in 2 K$^+$ from outside the cell
What provides the energy for the pump?	Hydrolysis of adenosine triphosphate (ATP) into adenosine diphosphate (ADP)
What substances inhibit activity of the pump?	Ouabain and digitalis glycosides

INTRACELLULAR CONNECTIONS

What are the most common types of intracellular connections?	Tight junctions and gap junctions
What is the role of tight junctions?	Attachments between cells that may be intercellular pathways
What types of tight junctions are there?	1. Impermeable (*tight*) 2. Permeable (*leaky*)
What is the role of gap junctions?	Attachments between cells that permit intercellular pathways

OSMOSIS

Define osmolarity.	Concentration of osmotically active particles in a solution
Define osmosis.	Flow of solvent across a semipermeable membrane from a low solute concentration to that with a higher solute concentration

Define osmotic pressure.	Driving pressure for water to move the given concentration of osmotically active particles
How is osmotic pressure calculated?	Van't Hoff's law: $\pi = gCRT$ where π = osmotic pressure (mm Hg), g = number of particles in solution (osm/mol), C = concentration (mol/L), R = gas constant (0.082 L·atm/mol·K), T = absolute temperature (K)
What is oncotic pressure?	Osmotic pressure created by proteins (e.g., colloid osmotic pressure)
Define the reflection coefficient.	Factor that describes the ease by which a solute is able to permeate a membrane
What values are possible for the reflection coefficient?	Any value between 0 and 1
What value of reflection coefficient represents impermeability?	1

MEMBRANE POTENTIALS

Define diffusion potential.	Potential difference across a membrane due to a difference in ion concentration
What determines the size of the diffusion potential?	Size of the concentration gradient
Can a diffusion potential be generated if the membrane is not permeable to the ion?	No
What determines the sign of the diffusion potential?	Charge of the diffusing ion
True or False? Creation of the diffusion potential requires the movement of a significant number of ions.	False. Very few ions move so as to not change the concentration of the diffusing ion
What facilitates the movement of ions across the membrane?	Ion channels
What does the conductance of an ion channel depend on?	Probability that the channel is open

General Physiology

What controls the opening and closing of ion channels?	Gates
What types of gates are there?	Voltage and ligand
What controls the opening and closing of voltage-gated channels?	Membrane potential
What controls the opening and closing of ligand-gated channels?	1. Hormones 2. Second messengers 3. Neurotransmitters
Define equilibrium potential (E).	Diffusion potential that exactly opposes diffusion caused by concentration difference
What is it called when the electrical and chemical driving forces of an ion are equally opposed?	Electrochemical equilibrium
What equation is used to calculate equilibrium potentials?	Nernst equation: $$E = -2.3 \frac{RT}{zF} \log_{10} \frac{[C_i]}{C_e}$$ where E = equilibrium potential (mV), $2.3\ RT/F$ = constants (60 mV at 37°C), z = charge on the ion, C_i = intracellular concentration (mM), C_e = extraceullular concentration (mM)
What are the approximate values of the equilibrium potential for the following ions in nerves and muscles:	
Na^+	+65 mV
K^+	−85 mV
Ca^{2+}	+120 mV
Cl^-	−85 mV
What is the resting membrane potential?	Sum of the diffusion potentials due to the concentration differences
Define depolarization.	Membrane potential becomes less negative (i.e., −60 → −40)
Define hyperpolarization.	Membrane potential becomes more negative (i.e., −60 → −90)
What is a cell called if it is capable of producing an action potential?	Excitable
What do action potentials consist of?	Rapid depolarization and repolarization

What are the unique characteristics of action potentials?	1. Stereotypical size and shape 2. Propagation 3. All-or-none events
What is the point after which an action potential is inevitable?	Threshold
What does inward or outward current charge refer to?	Movement of positive charge into (inward) and out of (outward) the cell
What is the resting membrane potential for nerve cells?	–70 mV
What ion is primarily responsible for the resting membrane potential of nerve cells?	K^+
What signals the activation gates of the Na^+ channels?	Depolarization of the cell membrane
What ion is responsible for the upstroke of the action potential in nerve cells?	Na^+
Define overshoot.	Peak of the action potential at which the membrane potential is positive
What signals the inactivation gates of the Na^+ channels?	Depolarization of the cell membrane
How can depolarization cause both activation and deactivation of Na^+ channel gates?	Inactivation occurs more slowly than activation
What causes repolarization of the cell membrane?	Closure of Na^+ channels and opening of K^+ channels
What signals the opening of K^+ channels?	Depolarization of the cell membrane
In what direction do K^+ ions move during repolarization?	Out of the cell (e.g., outward current)
What is the hyperpolarization of the afterpotential called?	Undershoot
What causes undershoot?	K^+ channels stay open after the Na^+ channels close
What types of refractory periods are there?	1. Absolute 2. Relative
In what type of refractory period can an action potential not be elicited?	Absolute refractory period
With what does the duration of the absolute refractory period coincide?	Nearly the entire length of the action potential

What produces the absolute refractory period?	While the inactivation gates of the Na^+ channels are closed, no action potential can be produced
How can an action potential be elicited during the relative refractory period?	With a larger than usual stimulus to produce a larger inward current
What is the duration of the relative refractory period?	Starts at the end of the absolute refractory period and lasts until the resting membrane potential is reached
Define accommodation.	Cell membrane is held at a depolarization level above threshold, but no action potentials are generated
How do action potentials propagate?	By spread of local currents to adjacent areas
How can conduction velocity be increased?	Increased fiber diameter and myelination
What type of conduction do myelinated axons demonstrate?	Saltatory
What type of a substance is myelin?	Insulating
Where are action potentials generated in myelinated axons?	Nodes of Ranvier
Describe salutatory conduction.	Depolarization jumps from one node of Ranvier to the next since current is unable to flow through myelin
Which has faster conduction, myelinated or unmyelinated fibers?	Myelinated fibers (conduction is up to 50 times faster)

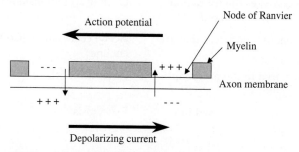

Figure 1.1 Saltatory Conduction.

NEUROMUSCULAR AND SYNAPTIC TRANSMISSION

What effect do action potentials have on presynaptic cells?	They depolarize the presynaptic terminal
Which ion is responsible for the presynaptic terminal's response to an action potential?	Ca^{2+}
What does depolarization of the presynaptic terminal lead to?	Release of neurotransmitter
Where is the neurotransmitter released into?	Synaptic cleft
What do neurotransmitters bind to?	Receptors on the cell membrane of the postsynaptic (most common) and/or presynaptic cell
What action do neurotransmitters have on postsynaptic cells?	Binding to their receptor causes a change in the membrane's permeability to specific ions and the membrane potential
What type of polarization do inhibitory neurotransmitters demonstrate?	Hyperpolarize postsynaptic membrane
What type of polarization do excitatory neurotransmitters demonstrate?	Depolarize postsynaptic membrane
What is the neuromuscular junction (NMJ)?	Synapse between an axon from a motoneuron and skeletal muscle
What is the neurotransmitter in the NMJ?	Acetylcholine (ACh)
What type of receptor is on the postsynaptic membrane?	Nicotinic receptor
What enzyme catalyzes the formation of ACh in the presynaptic terminal?	Choline acetyltransferase
What are the constituent pieces that are used to produce ACh?	Acetyl coenzyme A (CoA) and choline
Where is ACh stored in the presynaptic terminal?	Synaptic vesicles
By what process is ACh released into the synaptic cleft?	Exocytosis

What type of channel is the nicotinic receptor on the muscle end plate (postsynaptic membrane)?	Ligand-gated channel
What ion(s) permeability changes with binding of ACh to its receptor on the motor end plate?	Na^+ (primarily) and K^+
How does depolarization of the motor end-plate spread?	Local currents cause depolarization in adjacent muscle
What is it called when the many miniature end-plate potentials combine to produce the final end plate potential?	Summation
What types of summation are there?	1. Spatial summation 2. Temporal summation
Define spatial summation.	Two excitatory inputs from different inputs arrive simultaneously at the postsynaptic terminal and produce a greater depolarization
Define temporal summation.	Two excitatory inputs from the same input arrive back-to-back at the postsynaptic terminal and produce a stepwise increase in depolarization due to their overlap

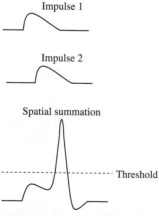

Figure 1.2 Spatial Summation.

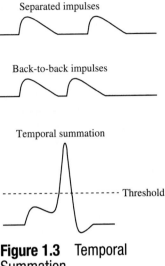

Figure 1.3 Temporal Summation.

What degrades the ACh in the synaptic cleft and on the motor end plate?	Acetylcholinesterase (AChE)
Why is degradation of ACh in the synaptic cleft and on the motor end plate necessary?	Permanent binding would cause permanent depolarization of the motor end plate and maintain muscle contraction (e.g., tetany)
What is it called when there is a single presynaptic element for each postsynaptic element?	One-to-one synapse
What is it called when there are many presynaptic elements for each postsynaptic element?	Many-to-one synapse
What happens when a many-to-one synapse receives both excitatory and inhibitory signals?	The postsynaptic cell integrates all of the signals and may or may not fire an action potential (AP)
What must be reached for the postsynaptic cell to fire an AP?	Threshold
What are the excitatory and inhibitory signals called in a many-to-one synapse?	Excitatory postsynaptic potential (EPSP) and inhibitory postsynaptic potential (IPSP)
What does an EPSP cause?	Opening of Na^+ and K^+ channels, which leads to depolarization

What does an IPSP cause?	Opening of Cl⁻ channels, which leads to hyperpolarization
What neurotransmitters are classically considered excitatory?	1. Ach 2. Norepinephrine (NE) 3. Epinephrine (Epi) 4. Dopamine 5. Glutamate 6. Serotonin
What neurotransmitters are classically considered inhibitory?	1. γ-Aminobutyric acid (GABA) 2. Glycine
What determines if a neurotransmitter is excitatory or inhibitory?	The neuron receptor subtype that it binds to
Diagram the synthetic pathway for dopamine, NE, and Epi.	Tyrosine ↓ *tyrosine hydroxylase* L-Dopa ↓ *dopa decarboxylase* Dopamine ↓ *dopamine β-hydroxylase* Norepinephrine ↓ *phenylethanolamine-N-methyltransferase* Epinephrine
Where is NE primarily released?	Postganglionic sympathetic neurons
What receptors does NE bind to?	α- or β-receptors
How is NE removed from the synaptic cleft?	1. Reuptake 2. Metabolized by monoamine oxidase (MAO) 3. Metabolized by catechol-*O*-methyltransferase (COMT)
Where is Epi mainly secreted?	Adrenal medulla
Where is dopamine mainly released?	Hypothalamus
Where is dopamine predominantly located?	Midbrain neurons (~80% is located in the basal ganglia)
What metabolizes dopamine?	MAO and COMT
What occurs when dopamine binds to the following receptors:	
D_1 receptor	Activates adenylate cyclase via a G_s protein
D_2 receptor	Inhibits adenylate cyclase via a G_i protein

What is serotonin derived from?	Tryptophan
What is serotonin converted into by the pineal gland?	Melatonin
Where are high concentrations of serotonin found?	Brain stem
What is histamine derived from?	Histadine
Where is histamine present as a neurotransmitter?	Hypothalamus
Which excitatory neurotransmitter is most prevalent in the brain?	Glutamate
What type of receptor does glutamate bind to?	Kainate receptor
What is the function of the glutamate receptor?	Na^+ and K^+ ion channel
What is GABA synthesized from?	Glutamate
What catalyzes GABA synthesis?	Glutamate decarboxylase
What occurs when GABA binds to the following receptors:	
$GABA_A$ receptor	Increases Cl^- conductance
$GABA_B$ receptor	Increases K^+ conductance
Which GABA receptor is the site of action for barbiturates and benzodiazepines?	$GABA_A$ receptor
Where is glycine primarily found?	Spinal cord and brain stem
What effect does glycine cause when it binds to its receptor?	Increases Cl^- conductance
What are the actions of the following agents on the neuromuscular junction:	
Botulinus toxin	Blocks release of ACh from the presynaptic terminal
Curare	Competes with ACh for receptors on the motor end plate
Hemicholinium	Blocks reuptake of choline into the presynaptic terminal
Neostigmine	Inhibits acetylcholinesterase

What are the effects on neuromuscular transmission of the following agents:	
Botulinus toxin	Total blockade of neuromuscular transmission
Curare	Decreases the size of the EPP
Hemicholinium	Prolongs and enhances the action of ACh at the muscle end plate
Neostigmine	Depletes the ACh store of the presynaptic terminal
What is the pathophysiologic basis of myasthenia gravis?	Presence of ACh receptor antibodies decreases the number of binding sites on the muscle end plate
What is the pathophysiologic basis of Parkinson's disease?	Degeneration of dopaminergic neurons in the substantia nigra and pars compacta that utilize D_2 receptors
What receptor type may be more prevalent in schizophrenics?	D_2 receptors

SKELETAL AND SMOOTH MUSCLE

What are the components of a skeletal muscle fiber?	1. Bundles of myofibrils 2. Surrounding sarcoplasmic reticulum (SR) 3. Transverse (T) tubules
What is a sarcomere?	Myofibril unit
Identify the labeled items in the following figure:	1. Sarcomere 2. I band 3. M line 4. Thin filament 5. H band 6. A band 7. Thick filament 8. Z line

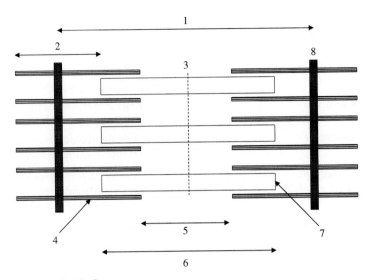

Figure 1.4 Sarcomere.

How are the filaments in a sarcomere arranged in skeletal muscle?	Longitudinally
What does the arrangement of sarcomeres produce on a macroscopic scale in skeletal muscle?	Striation pattern
What are thick filaments made up of?	Myosin
What are the components of myosin?	Six polypeptide chains: two heavy chains and four light chains
What are thin filaments made up of?	1. Actin 2. Tropomyosin 3. Troponin
What are the components of troponin and what are their functions?	Troponin C: Binds Ca^{2+} and allows interaction of actin and myosin Troponin I: Inhibits interaction of actin and myosin Troponin T: Attaches troponin to tropomyosin Remember: C = Ca^{2+}, I = inhibits, T = troponin
What are the functions of the SR?	Stores and releases Ca^{2+} in excitation-contraction coupling and maintains low intracellular $[Ca^{2+}]$

How are the SR and T tubule connected?	At terminal cisternae
How is Ca^{2+} stored in the SR?	Loosely bound to calsequestrin
How is Ca^{2+} transported into the SR?	Ca^{2+}-ATPase pump on the SR membrane
How does Ca^{2+} exit the SR?	Via the ryanodine receptor (Ca^{2+} channel)
What are the functions of T tubules?	Carry depolarization from the muscle cell membrane into the myofibrils
Describe the steps in excitation-contraction coupling that permit the cross-bridge cycle for muscle contraction.	Action potential occurs ↓ T tubules depolarize ↓ Ca^{2+} channels open on SR membrane ↓ Intracellular $[CA^{2+}]$ increases ↓ Ca^{2+} bind to troponin C ↓ Troponin undergoes conformational change moving tropomyosin from the myosin-binding site on actin ↓ Cross-bridge cycle occurs
Describe the steps in the cross-bridge cycle that produces muscle contraction.	Myosin has no ATP bound and is tightly attached to actin ↓ ATP binds to myosin causing a conformational change that releases actin ↓ Hydrolysis of ATP into ADP and inorganic phosphate (Pi) displace myosin toward the plus end of actin ↓ Myosin binds to a new site on actin, which generates the power for contraction ↓ ADP is released, returning myosin to its tightly bound state

What limits the cross-bridge cycle?	Ca^{2+} being bound to troponin C and exposing the myosin-binding sites on actin
What causes muscle relaxation?	Reuptake of Ca^{2+} by the SR
What would occur if the intracellular $[Ca^{2+}]$ remained high in a skeletal muscle cell?	The muscle would not be able to relax (e.g., tetanus)
What is an isometric contraction?	Force is generated without the muscle
What is an isotonic contraction?	Muscle fiber shortens at a constant afterload
What is afterload?	Load that the muscle contracts against
What is passive tension?	Tension developed by the muscle as it is stretched to different lengths
What is total tension?	Tension developed by the muscle when it contracts at different lengths
What is active tension?	The difference between total tension and passive tension
What is active tension proportional to?	The number of cross-bridges formed in the muscle
When is tension maximum?	When there is maximum overlap of thick and thin filaments permitting the most cross-bridges to form
Why does active tension decrease when the muscle is stretched maximally?	The number of cross-bridges that can form again are decreased compared to maximum
What happens to the velocity of shortening in a muscle as the afterload increases?	Decreases
What are the components of smooth muscle?	1. Thick filaments 2. Thin filaments
Are there sarcomeres in smooth muscle?	No
Is there troponin in smooth muscle?	No
What is the consequence of this?	Smooth muscle does not have striations
What are the types of smooth muscle?	1. Multiunit 2. Single-unit 3. Vascular
Which type is most common?	Single-unit

Which type has a high degree of electrical coupling between cells?	Single-unit
What type has spontaneous (*pacemaker*) activity?	Single-unit
What type is densely innervated?	Multiunit
Do multiunit smooth muscle units coordinate?	No, they behave separately
What properties does vascular smooth muscle exhibit?	Mix of multiunit and single-unit properties
What regulates excitation-contraction coupling in smooth muscle?	Ca^{2+} (there is no troponin!)
What are the mechanisms by which intracellular $[Ca^{2+}]$ can be increased in smooth muscle?	1. Depolarization of the cell membrane opens voltage-gated Ca^{2+} channels 2. SR may release additional Ca^{2+} with depolarization 3. SR can be stimulated by hormones and neurotransmitters to release Ca^{2+} via IP_3-gated Ca^{2+} channels
Describe excitation-contraction coupling in smooth muscle.	Intracellular $[Ca^{2+}]$ increases ↓ Ca^{2+} binds to calmodulin activating myosin light-chain kinase ↓ Myosin is phosphorylated and binds to actin ↓ Shortening occurs
What causes relaxation in smooth muscle?	Dephosphorylation of myosin

CHAPTER 2

Neurophysiology

AUTONOMIC NERVOUS SYSTEM

What are the components of the body's nervous system?	1. Somatic nervous system 2. Autonomic nervous system (ANS)
What is the function of the somatic nervous system?	Innervates skeletal muscle
What are the constituent parts of the ANS?	1. Enteric nervous system 2. Parasympathetic nervous system (PNS) 3. Sympathetic nervous system (SNS)
What is the function of the ANS?	Innervates and regulates cardiac muscle, smooth muscle, and glands
Where does the PNS originate?	1. Nuclei of cranial nerves (CN) III, VII, IX, and X 2. Spinal cord segments S2-S4 (e.g., craniosacral)
Where does the SNS originate?	Spinal cord segments T1-L3 (e.g., thoracolumbar)
What types of neurons are present in both the PNS and SNS and what are their corresponding neurotransmitters?	1. Adrenergic neuron; norepinephrine (NE) neurotransmitter 2. Cholinergic neuron; acetylcholinesterase (ACh) neurotransmitter
What type of neuron is unique to the PNS and what is its corresponding neurotransmitter?	Peptidergic neuron; peptide neurotransmitter (e.g., vasoactive inhibitory peptide or substance P)

Describe the axon length for the PNS and SNS in the following locations:	PNS	SNS
Preganglionic nerve axon	Long	Short
Postganglionic nerve axon	Short	Long

What receptor type(s) and neurotransmitter(s) are present in the preganglion fibers of the PNS and SNS?	Both use nicotinic receptors and ACh is the neurotransmitter
What is unique about the adrenal medulla?	It is a specialized SNS ganglion, where preganglionic fibers synapse directly onto the effector organ (chromaffin cells)
What does stimulation of chromaffin cells produce?	Secretion of epinephrine (Epi) (80%) and NE (20%) into the circulation
What receptor type(s) and neurotransmitter(s) are present in the effector organs of the PNS and SNS?	PNS: all receptors are muscarinic and ACh is the neurotransmitter SNS: α_1, α_2, β_1, β_2, and muscarinic receptors (in sweat glands) and NE is the neurotransmitter (except for muscarinic receptors where ACh is the neurotransmitter)
What are the SNS receptor types in the following locations and what are their effects:	
Eye	α_1: Constricts radial muscle and dilates iris β_1: Relaxes ciliary muscle
Bladder	β_2: Relaxes bladder wall α_1: Constricts bladder sphincter
Bronchioles	β_2: Dilates bronchiolar smooth muscle
Adipose tissue	β_1: Increase lipolysis
Gastrointestinal tract	α_1: Constricts gastrointestinal sphincters α_2: Decrease motility β_2: Decrease motility
Heart	β_1: Increase heart rate, contractility, and atrioventricular (AV) node conduction
Kidney	β_1: Increase renin secretion
Male sex organs	α: Ejaculation
Sweat glands	Muscarinic: Increase sweat production
Vascular smooth muscle	α_1: Constricts blood vessels in the skin and splanchnic circulation β_2: Dilates blood vessels in skeletal muscle

Neurophysiology

For which of the above locations are there no corresponding PNS muscarinic receptors?	1. Fat cells 2. Kidney 3. Sweat glands 4. Vascular smooth muscle
What is the mechanism of action for the following receptor types:	
α_1	Inositol 1, 4, 5-triphosphate (IP_3) formation and increased intracellular $[Ca^{2+}]$
α_2	Adenylate cyclase inhibition and decreased cyclic adenosine monophosphate (cAMP)
β_1	Adenylate cyclase activation and increased cAMP
β_2	Adenylate cyclase activation and increased cAMP
Muscarinic	Ion channel for Na^+ and K^+
Nicotinic	1. Heart (sinoatrial [SA] node): Adenylate cyclase inhibition 2. Smooth muscle and glands: IP_3 formation and increased intracellular $[Ca^{2+}]$
What is the effect on the ANS of the following agents:	
ACh	Nicotinic and muscarinic agonist
Albuterol	β_2 agonist
Atropine	Muscarinic antagonist
Butoxamine	β_2 antagonist
Carbachol	Nicotinic and muscarinic agonist
Clonidine	α_2 agonist
Curare	Nicotinic antagonist
Dobutamine	β_1 agonist
Hexamethonium	Nicotinic antagonist (ganglion only)
Isoproterenol	β_1 and β_2 agonist
Metoprolol (at therapeutic doses)	β_1 antagonist
Muscarine	Muscarinic agonist
Nicotine	Nicotinic agonist
NE	α_1 and β_1 agonist
Phenoxybenzamine	α antagonist
Phentolamine	α antagonist
Phenylephrine	α_1 agonist
Prazosin	α_1 antagonist

Propranolol	β_1 and β_2 antagonist
Yohimbine	α_2 antagonist
What autonomic centers are located in the following areas:	
Medulla	1. Respiratory 2. Swallowing, coughing, and vomiting 3. Vasomotor
Pons	Pneumotaxic
Midbrain	Micturition
Hypothalamus	1. Regulation of food and liquid intake 2. Temperature regulation

SENSORY SYSTEMS

What is a sensory receptor?	Specialized cell that transduce environmental signals into neural ones
What types of cells are usually sensory receptors?	Epithelial cells and neurons
What is a receptive field?	Area on the body that changes the firing rate of its sensory neuron when stimulated
What is the field called if it increases the firing rate?	Excitatory
What is the field called if it decreases the firing rate?	Inhibitory
How are sensory neurons classified?	By diameter (roman numerals I–IV) and conduction velocity (A and C)
What types of fibers are generally classified by their conduction velocity?	Sensory nerve fibers
Give the relative size of the following sensory neuron classifications:	
I	Largest
II	Medium
III	Small
IV	Smallest

Neurophysiology

Give the relative conduction velocity of the following sensory neuron classifications:	
Aα	Fastest
Aβ	Medium
Aδ	Medium
C	Slowest
Describe what occurs in sensory transduction.	Stimulus acts on sensory receptor ↓ Sensory receptor opens ion channels ↓ Change in membrane potential ↓ Receptor potential generated
What direction does the current usually flow when sensory receptor channels open?	Inward, depolarizing the cell
What is an exception to this flow direction?	Photoreceptors: stimulation decreases inward current and hyperpolarizes the membrane
What effect does size of the stimulus have on the receptor potential generated?	Larger stimuli create larger receptor potentials (e.g., graded potential)
What types of adaptation do sensory receptors exhibit?	Tonic (slow) and phasic (rapid)
Which type detects onset and offset of a stimulus?	Phasic
What type responds repetitively to prolonged stimuli?	Tonic
What type detects steady stimuli?	Tonic
What happens to action potential frequency in phasic receptors with constant stimulation?	Decreases
What is the location of the following structures:	
First-order neurons	Dorsal root or spinal cord ganglia
Second-order neurons	Spinal cord or brain stem
Third-order neurons	Thalamus (relay nuclei)
Fourth-order neurons	Cerebral cortex
At what level does sensory information cross the midline?	Second-order neurons (e.g., before reaching the thalamus)
What structure allows the sensory information to cross the midline?	Relay nucleus
What sensations are perceived by the somatosensory system?	1. Touch 2. Movement

	3. Temperature
	4. Pain
What pathways does the somatosensory system use?	Dorsal-column system and anterolateral system
What sensations are detected by the following systems:	
Dorsal-column system	1. Fine touch
	2. Pressure
	3. Two-point discrimination
	4. Vibration
	5. Proprioception
Anterolateral system	1. Temperature
	2. Pain
	3. Light touch
What type of nerve fiber is prevalent in the following systems:	
Dorsal-column system	Group II
Anterolateral system	Groups III and IV
What path does sensory information take in the dorsal-column system?	Receptors with cell bodies in the dorsal root receive stimulus ↓ Signal ascends to the nucleus gracilis and nucleus cuneatus in the medulla ↓ Signal crosses the midline and enters contralateral thalamus ↓ Signal sent to somatosensory cortex
What path does sensory information take in the anterolateral system?	Stimulus received by receptors in the periphery ↓ Signal crosses the midline and enters the anterolateral quadrant of the spinal cord ↓ Signal ascends to contralateral thalamus ↓ Signal sent to somatosensory cortex
What is another name for the sensory cortex?	Sensory homunculus
What are the types of mechanoreceptors that detect touch and pressure?	1. Meissner's corpuscle
	2. Merkel's disk

Neurophysiology

	3. Pacinian corpuscle
	4. Ruffini's corpuscle
What sensation is encoded by the following mechanoreceptors:	
Meissner's corpuscle	Velocity
Merkel's disk	Location
Pacinian corpuscle	Vibration and tapping
Ruffini's corpuscle	Pressure
Which types of mechanoreceptors demonstrate phasic adaptation?	Meissner's and Pacinian corpuscles
Which types of mechanoreceptors demonstrate tonic adaptation?	Merkel's disks and Ruffini's corpuscles
What is nociception?	Detection and perception of noxious stimuli (e.g., pain)
Where are pain receptors located?	1. Skin 2. Muscle 3. Viscera
What types of receptors detect pain?	No specialized receptors, instead pain is detected by free nerve endings
How is visceral pain perceived by the body?	Referred to the skin in dermatomal fashion
What fibers carry fast pain signals?	Group III
What fibers carry slow pain signals?	C fibers
What type of pain has rapid onset and offset?	Fast pain
What type of pain is poorly localized?	Slow pain
Opiates inhibit the release of what neurotransmitter for nociception?	Substance P

VISION

What is the function of the lens?	Focuses light onto the retina
What is the condition called when the curvature of the lens is not uniform?	Astigmatism
What conditions are described by the following:	
Lens focuses light onto the retina	Emmetropia (this is normal)

Lens focuses light in front of the retina	Myopia (nearsighted)
Lens focuses light behind the retina	Hyperopia (farsighted)
What is accommodation?	Focusing of the light by the lens
How does accommodation occur?	The smooth muscle of the ciliary body contracts to focus the lens and the pupils constrict
Which nerve is primarily involved in accommodation?	CN III
What is presbyopia?	Loss of the ability of lens to accommodate with age
Identify the labeled cell types of the retina in the image below?	1. Pigment epithelial cells 2. Receptors (rods and cones) 3. Bipolar cells 4. Horizontal cells 5. Amacrine cells 6. Ganglion cells

What are the functions of the following cell types:

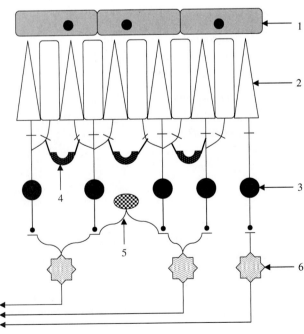

Figure 2.1 Retina.

Neurophysiology

Pigment epithelial cells	1. Prevent scatter of light by absorbing stray light 2. Convert 11-*cis* retinal to all-*trans* retinal (photoisomerization)
Horizontal cells	Form local circuits with bipolar cells in the outer plexiform layer
Amacrine cells	Form local circuits with bipolar cells in the inner plexiform layer

Identify the following characteristics of rods and cones (e.g., photoreceptors):	Cones	Rods
Acuity	High	Low
Type of light sensitive to	High-intensity	Low-intensity
Color vision	Present	Absent
Dark adaptation	Fast	Slow

Why do rods have lower acuity compared to cones?	Many rods synapse on a single bipolar cell, whereas only a few cones synapse on a single bipolar cell
What is the fovea?	Area on the retina where acuity is the highest
Are there rods present in the fovea?	No
What is the relationship of cones to bipolar cells in the fovea?	1:1
How is an image formed on the retina?	The image is inverted and reversed with respect to the object
What is the blind spot?	Location of the optic disk at approximately 15° off center where no rods or cones are present
What is the optic disk?	Location where the axons from the ganglion cells converge to form the optic nerve
What is the neural pathway of vision?	Signals from the axons of the ganglion cells in the retina pass caudally in the optic nerve into the optic tract and enter the lateral geniculate body (LGB) in the thalamus. Fibers from the LGB form the geniculocalcarine tract that travel to the primary visual cortex in the occipital lobe
What is the optic chiasm?	Location where some of the nerve fibers cross the midline to form the contralateral optic tract
Which hemiretina: nasal or temporal, crosses the midline in the optic chiasm?	Nasal hemiretina

Which side of the brain does fibers temporal to the fovea travel?	Ipsilateral
Where do the nerve fibers that form the right optic tract originate?	Right temporal hemiretina and left nasal hemiretina
Describe the visual defect present with the following lesions in the optic pathway:	
Right optic-nerve transection	Right eye is totally blind
Optic chiasm transection	Bitemporal hemianopia
Bilateral lateral compression of optic chiasm	Binasal hemianopia
Left optic tract	Right hemianopia
Left Meyer's loop transection	Right upper quadrantanopia (think: "pie in the sky")
Transection of right-visual radiation to cuneus	Left lower quadrantanopia
Left geniculocalcarine-tract transection	Right hemianopia with macular sparing
Where is the primary visual cortex located?	In the Brodmann area 17, located principally on the sides of the calcarine fissure
For what part of the visual field are most of the optic fibers carrying information?	90% carry information from the central $30°$ of the visual field
What is rhodopsin?	Visual pigment that absorbs light rays in the photoreceptors
What are the components of rhodopsin?	1. Opsin (protein) 2. Retinal (light-absorbing vitamin A analogue)
How does rhodopsin differ in rods and cones?	Rods have one type of opsin. Cones have three types: red, green, and blue
What conformation is retinal normally in?	11-*cis* retinal
What happens when light hits the retina?	11-*cis* retinal is converted to all-*trans* retinal (photoisomerization)
What is required to regenerate 11-*cis* retinal?	Vitamin A
Describe phototransduction from the formation of all-*trans* retinal.	All-*trans* retinal formed ↓ Rhodopsin undergoes multiple spontaneous transformations ↓ Metarhodopsin II is formed (active form of rhodopsin)

Neurophysiology

$\downarrow$

Transducin (G protein) activated, which activates phosphodiesterase

$\downarrow$

Phosphodiesterase catalyzes conversion of cyclic guanosine monophosphate (cGMP) to 5'-GMP

$\downarrow$

cGMP levels decrease

$\downarrow$

Na^+ channels close, decreasing inward Na^+ current, and hyperpolarizing the receptor cell membrane

$\downarrow$

Decreased release of neurotransmitter

What is the consensual reflex?	A reflex whereby shining a light in one eye normally leads to constriction of both pupils
Define nystagmus.	It is a repetitive, tremor-like oscillating movement of the eyes

HEARING AND BALANCE

How is sound transmitted to the brain?	Sound waves are directed to the auditory canal by the outer ear $\downarrow$ Transformed by the tympanic membrane and auditory ossicles into movements of the footplate of the stapes $\downarrow$ Cause wave formation in the perilymph of the inner ear $\downarrow$ Hair cells on the organ of Corti receive the sound in the form of fluid waves and generate action potentials in the nerve fibers
What are the auditory ossicles?	The three bones located in the middle ear: malleus, incus, and stapes
Which nerve is primarily involved in hearing and balance?	CN VIII
What areas in the brain are involved in hearing?	Areas 41 and 42

What is the function of tympanic membrane?	It vibrates in response to pressure changes produced by sound waves on its external surface and imparts its motions to the manubrium of the malleus
What makes up the inner ear?	1. Bony labyrinth (a series of channels in the petrous portion of the temporal bone) 2. Membranous labyrinth (duplicates the shape of the bony channels)
What makes up the bony labyrinth?	1. Semicircular canals 2. Cochlea 3. Vestibule
What are the fluids in the inner ear and where are they located?	1. Perilymph is the fluid outside the ducts 2. Endolymph is the fluid inside the ducts
What ion is predominant in the inner ear fluids?	Perilymph: Na^+ Endolymph: K^+
What structures form the cochlea?	1. Basilar membrane 2. Scala tympani 3. Scala vestibuli
Identify the labeled structures of the inner ear in the figure below.	1. Scala vestibuli 2. Scala media 3. Scala tympani 4. Basilar membrane 5. Tectorial membrane 6. Organ of Corti

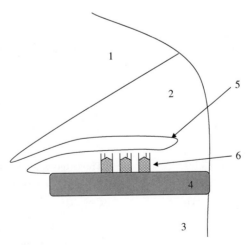

Figure 2.2 Inner Ear.

Neurophysiology

What are hair cells?	The receptor cells for auditory stimuli and acceleration (both angular and linear) of the head
Where are hair cells located?	1. Basilar membrane of the cochlea (audition) 2. Vestibular organ (acceleration)
What is the organ of Corti?	The structure located on the basilar membrane of the cochlea that contains the hair cells, which are the auditory receptors
How is sound transduced by the organ of Corti?	Sound waves cause vibration of the organ of Corti ↓ Cilia bend, which changes the K^+ conductance of the hair cell membrane ↓ Direction of bend causes depolarization or hyperpolarization ↓ Oscillating potential (cochlear microphonic potential) generated
What frequencies of sound are best detected in the following locations:	
Apex of the basilar membrane	Low frequencies
Base of the basilar membrane	High frequencies
What is the path of auditory nerve fibers?	Hair cells → lateral lemniscus → inferior colliculus → medial geniculate nucleus (MGN) of thalamus → auditory cortex
What is unique about the auditory nerve fiber's path?	Fibers may be crossed or uncrossed, which means that signals from both ears are present at all higher levels
How do you perform the following tests:	
Rinne's test	A 512 Hz tuning fork is struck and its handle placed on the patient's mastoid tip. The patient is asked if he or she hears a sound and if so indicates when the sound is no longer heard. At this point, the tines of the tuning fork are placed in front of the external auditory meatus of the same ear and the patient is asked if a sound is still heard. This is then repeated for the patient's other ear

Weber's test	A 512 Hz tuning fork is struck and the handle placed on the patient's forehead in the center. The patient indicates whether he or she hears or feels the sound in the right ear, the left ear, or in the middle of the forehead
How do the following characteristics differ in the Rinne's and Weber's test to distinguish conduction deafness from nerve deafness:	Rinne's Weber's
Normal hearing	Air conduction > bone conduction Hears equally on both sides
Conduction deafness	Bone conduction > air conduction Sound is louder in diseased ear
Nerve deafness	Bone conduction normal Sound is louder in normal ear
Which structures in the ear are concerned with equilibrium?	1. Semicircular canals 2. Utricle 3. Saccule of the inner ear
Which direction of acceleration do the following detect:	
Semicircular canal	Rotational
Utricle	Linear and horizontal
Saccule	Linear and vertical
What structure is direction of rotation relative to in the vestibular organ?	Kinocilium
What occurs when the stereocilia of the hair cell bend toward the kinocilium?	The hair cell depolarizes

TASTE AND OLFACTION

What is involved in the act of flavoring food and beverages?	1. Direct chemical stimulation of taste buds 2. Stimulation of olfactory receptors by vapors from food 3. Stimulation of chemical-sensitive and somatosensory free nerve endings of the trigeminal and other nerves in the mucous membranes of the oral and nasal cavities

What are taste buds?	Papillae on the surface of the tongue, as well as the pharynx
Are taste buds neurons?	No
What is the role of microvilli on taste buds?	Increases the surface area to bind taste chemicals
What types of papillae are there and where are they located?	1. Fungiform: anterior two-third of tongue 2. Circumvallate: posterior one-third of tongue 3. Foliate: posterior one-third of tongue
What are the four basic taste qualities?	1. Sweet 2. Salty 3. Sour 4. Bitter
What tastes do the various papillae discriminate?	Fungiform: salty and sweet Circumvallate: sour and bitter Foliate: sour and bitter
What transduction processes are involved with each of the basic taste qualities?	Sweet: Compounds bind to G-protein-coupled receptors resulting in decreased K^+ conductance Salty: Sodium channels in the apical membranes of taste receptor cells allow inward mobilization of Na^+ ions resulting in depolarization of the cell Sour: Acidic compounds cause increased H^+ concentration resulting in depolarization of receptor cells either by direct proton influx through sodium channels or by blocking conductance of pH-sensitive apical K^+ channels Bitter: Compounds often found in toxic substances. Receptor cells use both ligand-gated channels and G-protein-coupled receptors that result in influx of calcium ions from internal stores

Which CN innervate the following locations:

    Anterior two-third of tongue — CN VII (chorda tympani)

    Posterior one-third of tongue — CN IX (glossopharyngeal)

    Back of throat/epiglottis — CN X (vagus)

What is the pathway for taste? — CN VII, IX, and X enter the medulla and travel caudally to the solitary tract and synapse on second-order taste neurons in the solitary nucleus. Ascend from there ipsilaterally to ventral posteromedial nucleus of the thalamus and onto the taste cortex

How are olfactory receptors unique? — They are the only neurons that are replaced throughout a person's life

Which CN mediates olfaction? — CN I

What role does CN V play on olfaction? — It innervates the olfactory epithelium, which detects noxious or painful stimuli

What structure do the olfactory nerves pass through that when damaged can induce hyposmia or anosmia? — Cribriform plate

Which part of the brain are olfactory receptor neurons? — They are a direct part of the ipsilateral telencephalon and do not relay in the thalamus

What are the second-order olfactory neurons? — Mitral cells

Where does the output of the second-order olfactory neurons go? — To the prepiriform cortex

Describe the transduction of olfaction.

Olfactory receptor neurons bind odorant molecules
↓
Activate G proteins
↓
Activate adenylate cyclase
↓
Increase intracellular [cAMP]
↓
Na^+ channels opened
↓
Depolarization of receptor potential

MOTOR SYSTEMS

Define the following:

Motor unit	Single motoneuron and the muscle fibers it innervates
Small motoneuron	Motoneuron that innervates only a few muscle fibers
Large motoneuron	Motoneuron that innervates many muscle fibers
Motoneuron pool	Collection of motoneurons that innervate the muscle fibers of the same muscle

For small and large motoneurons identify the following characteristics:

	Small motoneuron	Large motoneuron
Force generation	Small	Large
Threshold level	Low	High
Firing time	Fast	Slow

How is muscle contraction increased?	By recruitment of additional motor units
What is the size principle?	As motor units are recruited, more motoneurons become involved, which generate more tension
What type of muscle sensors are there and what do they detect?	1. Muscle spindles: detect changes in muscle length (both dynamic and static) 2. Golgi tendon organs: detect muscle tension 3. Pacinian corpuscles: detect vibration 4. Free nerve endings: detect noxious stimuli
What types of muscle fibers are there?	Extrafusal and intrafusal
Which type generates the force for contraction?	Extrafusal fibers

What are the innervations for the following:

Extrafusal fibers	α-Motoneurons
Intrafusal fibers	γ-Motoneurons
How are intrafusal fibers arranged in relation to extrafusal?	They run in parallel, but not for the entire length of the muscle

What are muscle spindles?	Intrafusal fibers that have been encapsulated in sheaths and are connected in parallel with extrafusal fibers
What types of intrafusal fibers can be in muscle spindles?	Nuclear bag fibers and nuclear chain fibers
Which type is more numerous?	Nuclear chain fibers
What do the following fibers detect:	
Nuclear bag fibers	Dynamic changes in muscle length (e.g., velocity)
Nuclear chain fibers	Static changes in muscle length
What is the innervation for nuclear bag fibers?	Group Ia afferent fibers
What is the innervation for nuclear chain fibers?	Group II afferent fibers
What stimulus is detected by muscle spindles?	Muscle stretch (e.g., lengthened)
What is the response of nuclear bag fibers to stimulation?	Group Ia afferent fibers stimulate α-motoneurons in the spinal cord, which causes contraction in the muscle (e.g., shortening)
What role do γ-motoneurons play in the muscle fiber reflex?	Adjust sensitivity of the muscle spindle to produce the appropriate response during muscle contraction
What happens to the number of muscle spindles as movement becomes finer?	More spindles are recruited
What are the major muscle reflex types and how many synapses does each use?	1. Stretch reflex: monosynaptic 2. Golgi tendon reflex: disynaptic 3. Flexor-withdrawal reflex: polysynaptic
Describe the stretch reflex.	Muscle is stretched, stimulating group Ia afferent fibers ↓ Group Ia afferents synapse directly on α-motoneurons 1. Stimulation of α-motoneurons leads to contraction of the muscle that was stretched to return it to its original length 2. Synergistic muscles are activated and antagonistic muscles are inhibited

Neurophysiology

What happens to the stretch reflex if there is increased γ-motoneuron activity?	Exaggerates the reflex
Describe the Golgi tendon reflex.	Active muscle contraction stimulates group Ib afferent fibers in the Golgi tendon organs ↓ Group Ib afferents stimulate inhibitory interneurons in the spinal cord 1. α-Motoneurons inhibited, which cause relaxation of the contracting muscle 2. Antagonistic muscles are activated
Describe the flexor withdrawal reflex.	Pain stimulates groups II, III, and IV flexor reflex afferent fibers ↓ Afferents synapse via interneurons to multiple motoneurons in the spinal cord ↓ Ipsilateral side: Flexors are stimulated to contract and extensors are inhibited, pulling the side away from the stimulus Contralateral side: Flexors are inhibited and extensors are stimulated to maintain balance
What is the persistent neural activity that is left in the polysynaptic circuit of the flexor withdrawal reflex called?	Afterdischarge
What is the function of the afterdischarge?	Prevents the muscle from relaxing for some time
What are Renshaw cells?	Inhibitory interneurons in the ventral horn of the spinal cord
What are the pyramidal tracts?	1. Corticospinal tract 2. Corticobulbar tract
What are the extrapyramidal tracts (give both the name and the tract)?	1. Lateral vestibulospinal tract: Deiters' nucleus → ipsilateral motoneurons and interneurons 2. Medullary reticulospinal tract: Medullary reticular formation → interneurons in the intermediate gray area 3. Pontine reticulospinal tract: Nuclei in the pons → ventromedial spinal cord

	4. Rubrospinal tract: Red nucleus → lateral spinal cord interneurons 5. Tectospinal tract: Superior colliculus → cervical spinal cord
What does stimulation of the following tracts produce:	
Lateral vestibulospinal tract	Stimulates extensors and inhibits flexors
Medullary reticulospinal tract	Inhibits both (extensors > flexors)
Pontine reticulospinal tract	Stimulate both (extensors > flexors)
Rubrospinal tract	Stimulates flexors and inhibits extensors
Tectospinal tract	Controls neck muscles
What are the functions of the cerebellum?	1. Controls balance and eye movement (vestibulocerebellum) 2. Plans and initiates movement (pontocerebellum) 3. Controls rate, force, range, and direction of movement (spinocerebellum)
What are the layers of the cerebellar cortex from outside in?	1. Molecular layer 2. Purkinje cell layer 3. Granular layer
Where does the cerebellar output come from?	Purkinje cells
What type of output does the cerebellum produce?	Always inhibitory
What is the neurotransmitter for the cerebellar output?	γ-Aminobutyric acid (GABA)
What are the components of the basal ganglia?	1. Striatum 2. Globus pallidus 3. Subthalamic nuclei 4. Substantia nigra
What is the function of the basal ganglia?	Plans and executes smooth movements by modulating thalamic outflow to the motor cortex
How does the basal ganglia communicate with the thalamus and cerebral cortex?	1. Direct pathway 2. Indirect pathway
Which pathway has an overall inhibitory effect?	Indirect pathway
Which pathway has an overall excitatory effect?	Direct pathway

What is the neurotransmitter for communication between the striatum and substantia nigra?	Dopamine
What effect does dopamine have on the following:	
Direct pathway	Excitatory (D_1 receptors)
Indirect pathway	Inhibitory (D_2 receptors)
What is the effect of a lesion in the following locations:	
Striatum	Produces quick, continuous, and uncontrollable movements (e.g., chorea)
Globus pallidus	Produces inability to maintain postural support
Subthalamic nuclei	Produces wild, flinging movements (e.g., hemiballismus)
Substantia nigra	Produces cogwheel rigidity, tremor, and decreased voluntary movement
What disease produces signs similar to a lesion in the striatum?	Huntington's disease
What disease produces signs similar to a lesion in the substantia nigra?	Parkinson's disease
Where in the brain are the premotor cortex and the supplementary motor cortex located?	Area 6
What are the functions of the premotor cortex and supplementary motor cortex?	Generate a plan for movement and transfer it to the primary motor cortex
Which cortex is utilized in *mental rehearsal* of a movement?	Supplementary motor cortex
What area is the primary motor cortex?	Area 4
What is the function of the primary motor cortex?	Execution of movement
What is the somatotopical organization of the primary motor cortex known as?	Motor homunculus
What occurs when epileptic events occur in the primary motor cortex?	Jacksonian seizure (Jacksonian march), which is a seizure that spreads through the primary motor cortex in succession and affects the corresponding muscle groups

What occurs with transection of the spinal cord below the lesion?	1. Loss of voluntary movement 2. Loss of conscious sedation 3. Loss of reflexes (initially)
What occurs with time to the initial loss of reflexes?	Partial return or may progress to hyperreflexia
A transection above what level will produce the following:	
Loss of sympathetic tone to the heart	C7
Breathing cessation	C3
Death	C1
What is the effect due to lesions above the following locations:	
Lateral vestibular nucleus	Decerebrate rigidity
Red nucleus	Decorticate posturing with intact tonic neck reflexes
Pontine reticular formation (below the midbrain)	Decerebrate rigidity
Describe the posture of the following conditions:	
Decerebrate rigidity	1. Arms are adducted and rigidly extended at the elbows 2. Forearms are pronated 3. Wrists and fingers are flexed 4. Feet are plantarflexed
Decorticate rigidity	1. Arms are adducted 2. Elbows, wrists, and fingers are flexed 3. Legs are internally rotated 4. Feet are plantarflexed

CEREBRAL CORTEX FUNCTIONS

What do electroencephalogram (EEG) potentials measure?	Synaptic potentials in the pyramidal cells of the cerebral cortex
What EEG waves predominate in the following conditions:	
Alert and awake	Beta waves
Relaxing and awake	Alpha waves
Asleep	Slow waves

Neurophysiology

What is the rhythm that governs sleep called?	Circadian rhythm
Where is the rhythm thought to be controlled?	Suprachiasmatic nucleus of the hypothalamus
What is REM sleep?	Rapid eye movement sleep
What does an EEG look like when someone's in REM sleep?	As if they were awake (e.g., beta waves predominate)
What occurs during REM sleep?	1. Eye movements 2. Pupillary constriction 3. Dreaming 4. Loss of muscle tone 5. Penile erection (in males)
How often does REM sleep normally occur?	Every 90 minutes
What happens to the duration of REM sleep with age?	It decreases
What commonly prescribed medication has the same effect on REM sleep as age?	Benzodiazepines
What allows for the transfer of information between the hemispheres of the cerebral cortex?	Corpus collosum
In what hemisphere is language dominant?	Left hemisphere
Does handedness affect the dominant language hemisphere?	No
What aspects of language are dominant in the right hemisphere?	1. Body language 2. Facial expression 3. Intonation 4. Spatial tasks
Define aphasia.	Loss or impairment of the power to use or comprehend words
What type of aphasia is it when a person cannot understand written or spoken language?	Sensory aphasia
A lesion in what area will cause such an aphasia?	Wernicke's area
What type of aphasia is it when a person can understand language, but not write or speak?	Motor aphasia

A lesion in what area will cause such an aphasia?	Broca's area
How are short-term memories formed?	By changing synaptic connections
How are long-term memories formed?	By making structural changes in the nervous system (more stable)
Lesions to what will prevent formation of new long-term memories?	Bilateral lesions on the hippocampus

TEMPERATURE REGULATION

What structure regulates body temperature?	Anterior hypothalamus
How is body temperature regulated?	Core temperature is determined by temperature receptors on the skin and in the hypothalamus. This value is compared in the anterior hypothalamus with the set-point temperature. If the core temperature is off from the set-point, appropriate mechanisms are activated
What are the body's heat-producing mechanisms and how do they work?	1. Thyroid hormone: stimulates Na^+-K^+-ATPase to increase metabolic rate and produce heat 2. SNS: activates β receptors in brown fat to increase metabolic rate and produce heat 3. Shivering: posterior hypothalamus activates α- and γ-motoneurons causing muscle contraction and heat production
What are the body's mechanisms to get rid of excess heat and how do they work?	1. SNS: decreased tone to cutaneous blood vessels increases arteriovenous shunting of blood to venous plexuses close to the skin's surface and increases convection and radiation to produce heat loss (mediated by posterior hypothalamus) 2. Sweat glands: activated by sympathetic muscarinic receptors to produce heat loss by evaporation
What is a pyrogen?	Substance that increases the body's set-point temperature

How is fever mediated through a pyrogen?	Pyrogens increase interleukin-1 (IL-1) production ↓ IL-1 stimulates the anterior hypothalamus to produce prostaglandins ↓ Prostaglandins raise the set-point temperature
Define heat exhaustion.	Decreased blood pressure or syncope that develops from decreased blood volume due to excessive sweating
Define heat stroke.	Normal responses to increased ambient temperature are impaired and core temperature rises to the point of tissue damage
Define malignant hyperthermia.	Skeletal muscle produces excess heat through massive oxygen consumption and causes a rapid rise in body temperature
What commonly used agents can cause malignant hyperthermia?	Inhaled anesthetics
What agent is used to reverse malignant hyperthermia?	Dantrolene
Define hypothermia.	Normal responses to low ambient temperature are insufficient to maintain core temperature near the set-point

CHAPTER 3

Cardiovascular Physiology

VASCULATURE

How do arteries and veins differ in regards to the following:	Arteries	Veins
Function	Deliver oxygenated blood to tissues	Return blood to the heart
Wall thickness	Thick-walled (due to elastic tissue and smooth muscle)	Thin-walled
Pressures	High pressure	Low pressure
Volume contained	Stressed volume	Unstressed volume

Which contains the larger proportion of blood, arteries or veins?
Veins contain the highest proportion of the blood, the unstressed volume

What is Poiseuille's equation for resistance

$$R = \frac{8\eta l}{\pi r^4}$$

where R = resistance, η = viscosity of blood, l = length of the blood vessel, r = radius of the blood vessel

Which component of the vascular system is the site of the highest resistance?
The arterioles

How is resistance regulated?
Through the autonomic nervous system

On which arterioles are α_1-adrenergic receptors found?
1. Skin
2. Renal circulation
3. Splanchnic circulation

45

On which arterioles are β_2-adrenergic receptors found?	Skeletal muscle
What is unique about the capillary bed system of the vasculature? And what is its role?	It has the largest cross-sectional area and surface area, which allows for the efficient exchange of nutrients, water, and gases
What is unique about the pulmonary vasculature compared to the systemic vasculature?	Hypoxia causes vasoconstriction of the pulmonary vasculature. In most organs hypoxia causes vasodilation

HEMODYNAMICS

What is the equation for the velocity of blood flow?	$v = \dfrac{Q}{A}$ where v = velocity of blood flow (cm/s), Q = blood flow (mL/s), A = cross-sectional area (cm^2)
Why is velocity of blood flow in the aorta higher than that of the capillaries?	Velocity of blood flow is inversely proportional to cross-sectional area and the aorta has a relatively small cross-sectional area compared to that of the sum of all the capillaries
What famous law is the equation for blood flow analogous to?	Ohm's law: $I = V/R$
What factors change the resistance of the vasculature system proportionally?	Viscosity and length of the vessel
What is the major determinant of viscosity in the vascular system?	The hematocrit is mostly responsible for the viscosity within the vascular system
In what physiological states does viscosity increase?	1. Polycythemia 2. Hyperproteinemia 3. Hereditary spherocytosis
What happens to the resistance if the radius of the blood vessel is reduced by 50%?	Increased by a factor of 16 ($R \propto [1/r^4]$)
Where does the largest decrease in vascular pressure occur?	Across the arterioles, which is the site of greatest resistance in the system

Cardiovascular Physiology

What is the equation for capacitance?	$$C = \frac{V}{P}$$ where C = capacitance (mL/mm Hg), V = volume (mL), P = pressure (mm Hg)
Which has more capacitance, an artery or a vein? Why?	Capacitance is inversely related to elastance. Since arteries have more elastic tissue compared to veins, they have less capacitance
When is the systolic pressure measured?	After the heart contracts
When is diastolic pressure measured?	After the heart relaxes
How can you calculate the pulse pressure?	Pulse pressure = systolic − diastolic
What is the major determinant of pulse pressure?	Stroke volume (SV)
What happens to the pulse pressure when the capacitance decreases?	Increases
What is the mean arterial pressure (MAP)?	Average arterial pressure with respect to time
How can you determine MAP?	MAP = CO × TPR or MAP = ⅓ systolic + ⅔ diastolic Where CO = cardiac output, TPR = total peripheral resistance
Which is lower, venous pressure or atrial pressure?	Atrial pressure
How can the left atrial (LA) pressure be estimated?	Pulmonary capillary wedge pressure (PCWP)
What is used to measure LA pressure?	Swan-Ganz catheter

ELECTROPHYSIOLOGY—ELECTROCARDIOGRAM

Identify the labeled parts and intervals of the electrocardiogram (ECG) below.

1. P wave
2. QRS complex
3. T wave
4. PR interval
5. ST segment
6. QT interval

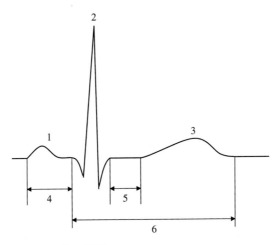

Figure 3.1 ECG.

Define what the following waves/
complexes from the ECG represent:

P wave	Atrial depolarization
QRS complex	Ventricular depolarization
T wave	Ventricular repolarization
When does atrial repolarization occur?	During the QRS complex, but it is masked by the larger signal from ventricular depolarization
When does conduction through the atrioventricular (AV) node take place?	During the PR interval
What happens during the QT interval?	Mechanical contraction of the ventricles
When might you see a U wave on an ECG?	Hypokalemia

What are the normal time values for
the following:

PR interval	0.12–0.2 second
QRS complex	0.12 second

What are the ECG changes associated
with the following types of heart blocks:

1° (first degree)	PR interval >0.20 second
2° (second degree) Mobitz type 1	PR intervals progressively increase from beat to beat until they are long enough that a beat is dropped
2° (second degree) Mobitz type 2	PR interval >0.20 second at a fixed interval with a fixed ratio of dropped beats
3° (third degree)	No relationship between P wave and QRS complex

What are the ECG changes associated with the following abnormal rhythms:

Atrial fibrillation — Irregularly spaced QRS complexes with intervening erratic spikes instead of P waves (e.g., irregularly irregular)

Atrial flutter — "Sawtooth" baseline

Ventricular fibrillation — Completely abnormal rhythm that has no recognizable waves or complexes

ELECTROPHYSIOLOGY—CARDIAC ACTION POTENTIAL

Conductance of what determines the resting membrane potential in cardiac muscle cells? — Conductance to K^+

What is the resting membrane potential of the nonpacemaker components of the heart? — ~90 mV, which is close to the K^+ equilibrium potential

What maintains the resting membrane potential? — Na^+-K^+-ATPase membrane protein

What does it mean when a membrane *depolarizes*? — There is an inward current that brings in positive charge into the cell

What does it mean when a membrane *hyperpolarizes* or *repolarizes*? — There is an outward current that removes positive charge from the cell

Diagram a cardiac action potential and label all of its phases and the currents that are responsible for them.

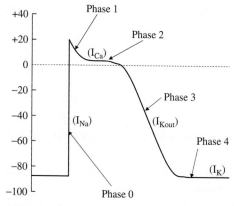

Figure 3.2 Cardiac Action Potential.

Describe what happens during each of the following phases of a cardiac action potential:

Phase 0 — Upstroke: caused by a rapid transient increase in Na^+ conductance, which allows an inward Na^+ current to depolarize the membrane (I_{Na})

Phase 1 — Brief initial repolarization: caused by an outward K^+ current, which is due to both chemical and electrical gradients; and decrease in inward Na^+ current due in part to a decrease in Na^+ conductance

Phase 2 — Plateau phase: caused by transient increase in Ca^{2+} conductance and an increase in K^+ conductance, which causes inward Ca^{2+} and outward K^+ currents (I_{Ca})

Phase 3 — Repolarization: caused by decrease in Ca^{2+} conductance and an increase in K^+ conductance resulting in a large outward K^+ current (I_{Kout})

Phase 4 — Resting membrane potential: caused by an equilibrium between outward and inward K^+ currents (I_K)

What determines the peak of the cardiac action potential? — Conductance to Na^+

ELECTROPHYSIOLOGY—PACEMAKER POTENTIAL

Where do pacemaker potentials occur normally?
1. Sinoatrial (SA) node
2. AV node
3. His-Purkinje systems

What is the normal pacemaker of the heart? — The SA node

What is the normal rate of pacemaker potentials? — 60–100 potentials/min

What are the latent pacemakers in the heart? — AV node and His-Purkinje system

When might latent pacemakers take over for the main pacemaker? — When the SA node is suppressed

Cardiovascular Physiology

What is unique about the pacemaker action potential?

It has an unstable resting potential that exhibits Phase 4 depolarization causing automaticity

What phases are not present in the pacemaker potential?

Phases 1 and 2 are not present

Diagram a cardiac pacemaker potential and label all of its phases and the currents that are responsible for them.

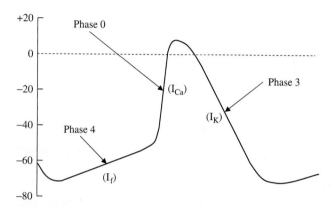

Figure 3.3 Pacemaker Potential.

Describe what happens during each of the following phases of the pacemaker potential:

Phase 0 — Upstroke: caused by an increase in Ca^{2+} conductance, which allows an inward Ca^{2+} current to depolarize the membrane toward the Ca^{2+} equilibrium potential (I_{Ca})

Phase 3 — Repolarization: caused by an increase in K^+ conductance resulting in an outward K^+ current (I_K)

Phase 4 — Slow depolarization: caused by an increase in Na^+ conductance, resulting in an inward Na^+ current (I_f)

What is the name of the current that accounts for pacemaker activity?	I_f
What turns on the current that accounts for the pacemaker activity?	Repolarization of the membrane potential in the preceding action potential turns on I_f

ELECTROPHYSIOLOGY—CONDUCTION AND EXCITABILITY

What is conduction velocity?	The rate at which an excitation impulse is able to spread throughout the cardiac tissue
What does conduction velocity depend on?	The magnitude of the inward current due to the influx of ions during Phase 0 of the cardiac action potential
Where is conduction velocity the fastest?	Purkinje system
Where is conduction velocity the slowest?	AV node
What does the difference in conduction time between the AV node and the Purkinje system allow the heart to do?	It allows for ventricular filling
What can happen if conduction velocity through the AV node is increased?	Ventricular filling can become compromised
What is a dromotropic effect?	Anything that produces a change in conduction velocity, primarily in the AV node
What type of dromotropic effect does the sympathetic nervous system produce?	Positive dromotropic effect
How does the sympathetic nervous system produce its dromotropic effect?	Increases conduction velocity through the AV node and decreases the PR interval by increasing the inward Ca^{2+} current
What is the receptor and neurotransmitter for the sympathetic effect?	β_1 receptor; norepinephrine (NE) is the neurotransmitter
What type of dromotropic effect does the parasympathetic nervous system produce?	Negative dromotropic effect
How does the parasympathetic system produce its dromotropic effect?	Decreases conduction velocity through the AV node and increases the PR interval by decreasing the inward Ca^{2+} current and increasing the outward K^+ current

Cardiovascular Physiology

What is the receptor and neurotransmitter for the parasympathetic effect?	Muscarinic receptor; acetylcholine (ACh) is the neurotransmitter

Diagram and identify the following in cardiac action potential:
1. Absolute refractory period (ARP)
2. Effective refractory period (ERP)
3. Relative refractory period (RRP

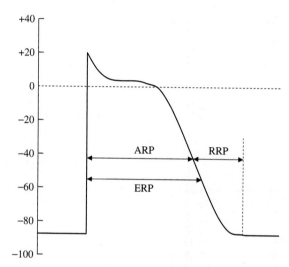

Figure 3.4 Refractory Periods.

What is excitability?	The ability of the cardiac muscle cells to initiate an action potential after being depolarized by an inward current
What is ARP?	The time during which no action potential can be initiated regardless of the magnitude of the inward current
What is ERP?	The time during which no conducted action potential can be produced
What is RRP?	The time during which an action potential can be initiated, but requires a larger inward current
What is the starting and ending point of the following refractory periods:	
ARP	Starts with the upstroke and ends after the plateau

ERP	Starts with the upstroke and ends slightly after the ARP
RRP	Starts right after the ARP and lasts until repolarization is complete

CARDIAC MUSCLE

What is a sarcomere?	Contractile unit of a myocardial cell
What is the role of the intercalated disk?	Maintains cell-to-cell cohesion
What is the role of the gap junction?	Provides a low-resistance path between cells, which allows for the rapid spread of action potentials
What is the role of the T tubule?	Carries action potentials into the myocardial cell
What is the role of the sarcoplasmic reticulum (SR)?	Stores and releases Ca^{2+} for myocardial cell excitation-contraction coupling
What occurs in excitation-contraction coupling within the myocardium once the action potential enters the cell via the T tubules?	Ca^{2+} enters the cell from the extracellular fluid (ECF), creating an inward Ca^{2+} current ↓ Influx of Ca^{2+} causes the SR to release its stores of Ca^{2+}, further increasing the intracellular $[Ca^{2+}]$ ↓ Ca^{2+} binds to troponin C molecules, which removes the inhibition of actin and myosin binding by moving tropomyosin out of the way ↓ Binding of actin and myosin allow the thick and thin filaments to slide past each other contracting the myocardial cell ↓ The SR reaccumulates the Ca^{2+}, which causes the myocardial cell to relax
What influences the amount of Ca^{2+} released by the SR during myocardial contraction?	Amount of Ca^{2+} stored in the SR and the size of the inward Ca^{2+} current
What is the magnitude of the contraction of a myocardial cell proportional to?	The intracellular $[Ca^{2+}]$
Is reuptake of Ca^{2+} by the SR during relaxation of the myocardium an active or passive process?	It's an active process mediated by Ca^{2+}-ATPase pump

What is cardiac oxygen consumption related to?	It is directly related to the amount of tension developed in the cardiac muscle
What are some factors that can increase cardiac oxygen consumption?	1. Increased afterload 2. Increased contractility 3. Increased heart rate (HR) 4. Hypertrophy of the cardiac muscle

CONTRACTILITY

What is contractility?	The ability of the myocardium to develop a force at a given muscle length
What is another term for contractility?	Inotropism
What can be used as an estimate of contractility?	Ejection fraction (EF)
What is the expression for EF?	$EF = \dfrac{SV}{EDV} \times 100\%$
What is the normal range of EF?	The normal range of EF ≈ 55–80%
What is a positive (negative) inotropic agent?	Anything that causes an increase (decrease) in contractility
What are some positive inotropes?	1. Increased HR 2. Sympathetic stimulation 3. Increased intracellular Ca^{2+} 4. Decreased extracellular Na^+ 5. Cardiac glycosides
What are some negative inotropes?	1. Parasympathetic stimulation 2. β_1-blockade 3. Heart failure 4. Acidosis 5. Hypoxia 6. Hypercapnea
How does sympathetic stimulation increase contractility?	Increases inward Ca^{2+} current during Phase 2 and phospholamban is phosphorylated to increase the activity of the Ca^{2+} pump of the SR providing more Ca^{2+} for release
What receptor does the sympathetic nervous system use to increase contractility?	β_1 receptors

What is the neurotransmitter for the sympathetic nervous system effect?	Catecholamines
How does parasympathetic stimulation decrease contractility?	Muscarinic receptors are stimulated by ACh to decrease the inward Ca^{2+} current during Phase 2
How do cardiac glycosides increase contractility?	Myocardial cell membrane Na^+-K^+-ATPase is inhibited, which diminishes the Na^+ gradient across the cell membrane. This decreased gradient causes the Na^+-Ca^{2+} exchange mechanism to increase intracellular Ca^{2+}

STARLING RELATIONSHIPS

What relationship is described by a Starling curve?	Change in SV and cardiac output (CO) that occurs due to changes in venous return or end-diastolic volume
What is the expression for SV?	SV = EDV − ESV where EDV = end-diastolic volume, ESV = end-systolic volume
What is the length-tension relationship in the ventricles?	Relates the length of the ventricular muscle cell to the force of contraction it generates
What factors govern the Starling/length-tension relationships?	1. Preload 2. Afterload 3. Sarcomere length 4. Velocity of contraction at a fixed muscle length
What is preload equivalent to?	Ventricular EDV
What effect do venodilators have on preload?	Decrease it
What are some things that can cause an increase in preload?	1. Increased blood volume 2. Sympathetic stimulation 3. Exercise (slight increase)
What is afterload equivalent to?	Diastolic arterial pressure
What is afterload proportional to?	Peripheral resistance
What effect do vasodilators have on afterload?	Decrease it

Cardiovascular Physiology

What occurs to the Starling curve under the following conditions:

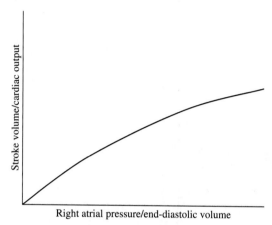

Figure 3.5 Starling Curve.

1. Increased contractility

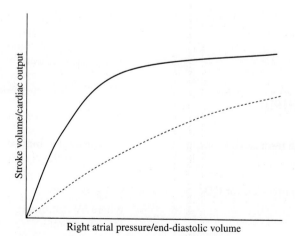

Figure 3.6 Increased Contractility.

2. Decreased contractility

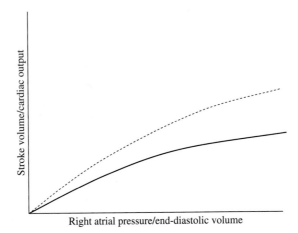

Figure 3.7 Decreased Contractility.

What can increase the contractility of the myocardium?	1. Digitalis 2. Sympathetic stimulation 3. Circulating catecholamines (epinephrine) 4. Abrupt increase in afterload (Anrep effect) 5. Increased HR (Bowditch effect)
What can decrease the contractility of the myocardium?	1. Pharmacologic depressants 2. Parasympathetic stimulation 3. Loss of myocardium 4. Congestive heart failure

CARDIAC OUTPUT

What does the term cardiac output (CO) mean?	It is the total volume of blood pumped by the ventricle per minute
What is the expression for CO?	$CO = SV \times HR$ where SV = stroke volume and HR = heart rate

Cardiovascular Physiology

How can CO be expressed through the equation for the velocity of blood flow?

$$CO = \frac{MAP - RA\,pressure}{TPR}$$

or

$$Q = \frac{\Delta P}{R}$$

where MAP = mean arterial pressure, RA = right atrial, TPR = total peripheral resistance, ΔP = pressure gradient (mm Hg), R = resistance (mm Hg/mL/min)

CARDIAC CYCLE

What are the steps of the cardiac cycle?

Atrial systole
↓
Isovolumetric ventricular contraction
↓
Rapid ventricular ejection
↓
Reduced ventricular ejection
↓
Isovolumetric ventricular relaxation
↓
Rapid ventricular filling
↓
Reduced ventricular filling
↓
Repeat . . .

Identify the labeled phases and events in the ventricular pressure-volume loop on the following page:

1. Ventricular filling
2. Isovolumetric contraction
3. Ventricular ejection
4. Isovolumetric relaxation
5. Mitral valve opens
6. Mitral valve closes
7. Aortic valve opens
8. Aortic valve closes

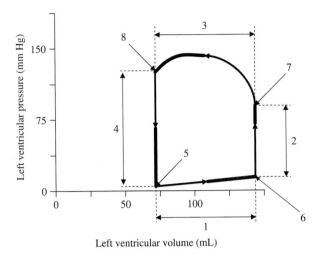

Figure 3.8 Pressure-Volume Loop.

What happens during each of the following phases of the cardiac cycle:

Isovolumetric contraction — Left ventricle (LV) is filled with blood from the left atrium (LA) and the ventricular muscle is relaxed. Upon excitation, the ventricle contracts and ventricular pressure increases, once LV pressure > LA pressure, the mitral valve closes. All valves are then closed and no blood is ejected, but the LV contraction continues and LV pressure increases

Ventricular ejection — When LV pressure > aorta pressure, the aortic valve opens and blood is ejected out of the LV and into the aorta

Isovolumetric relaxation — LV begins to relax and when LV pressure < aorta pressure, the aortic valve closes. All valves are closed at this point, but the ventricle continues to relax

Ventricular filling — When LV pressure < LA pressure, the mitral valve opens and the ventricle begins to fill

What is the volume of the ventricle before isovolumetric contraction? — EDV

What is the width of the pressure-volume loop a measurement of? — SV

Cardiovascular Physiology

What is the volume of the ventricle at the end of ventricular ejection?

ESV

What happens to the pressure-volume loop under the following conditions and what causes these changes?

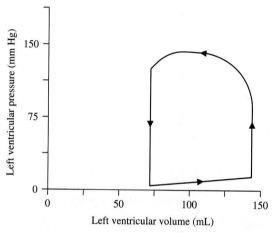

Figure 3.9 Blank Loop.

1. Increased preload

    Increased preload results from increased venous return and causes an increase in EDV. This increase causes an increase in SV (due to Frank-Starling relationship), which is reflected as an increased width of the loop

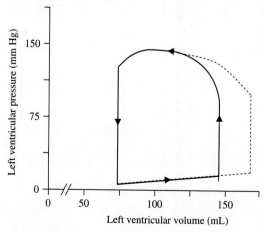

Figure 3.10 Increased Preload Curve.

2. Increased afterload

Increased afterload results from an increase in aortic pressure, which leads to a decrease in SV. This decreases the width of the loop

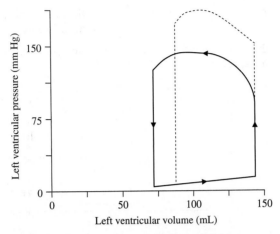

Figure 3.11 Increased Afterload Curve.

3. Increased contractility

Increased contractility leads to the ventricle developing greater tension during systole and increases the SV

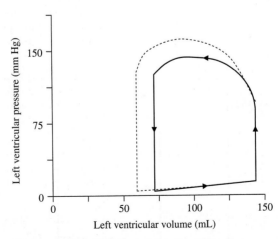

Figure 3.12 Increased Contractility Curve.

HEART SOUNDS AND MURMURS

What event does the first heart sound (S1) correspond to?	Mitral and tricuspid valve closure
What is a *split S1*?	When the mitral valve closes before the tricuspid valve
What event does the second heart sound (S2) correspond to?	Aortic and pulmonic valve closure
What are the potentially pathological heart sounds?	S3 and S4
In what instances are the heart sounds not pathological?	S3: normal in children S4: always pathological
What pathology are the abnormal heart sounds associated with?	S3: dilated congestive heart failure (CHF) S4: hypertrophic ventricle
When would you hear the abnormal heart sounds?	S3: at the end of the rapid ventricular filling phase S4: high atrial pressure/stiff ventricle and occurs between S2 and S1
Describe the murmurs associated with the following conditions. Also provide where are they best heard, and where they radiate to:	
Aortic stenosis	Ejection click followed by a midsystolic crescendo-decrescendo murmur at second right interspace and radiates to carotids and apex
Aortic regurgitation	Blowing early diastolic murmur at the aorta and left sternal border, apical diastolic rumble (Austin Flint murmur), and midsystolic flow murmur at the base
Mitral stenosis	Opening snap followed by a delayed rumbling diastolic murmur at apex
Mitral regurgitation	Holosystolic high-pitched blowing apical murmur at the left sternal border
Mitral valve prolapse (MVP)	Late systolic murmur with midsystolic click
Ventricular septal defect (VSD)	Holosystolic murmur over entire precordium and heard maximally at the fourth left interspace

Patent ductus arteriosus (PDA)	Continuous machine-like murmur that is loudest at second left interspace

ARTERIAL PRESSURE REGULATION

Through what reflex is the body able to quickly regulate the minute-to-minute arterial blood pressure?	Baroreceptor reflex
How is it mediated?	It is a neurally mediated negative feedback system
What are baroreceptors?	Stretch receptors
Where are they located?	In the carotid sinuses by the bifurcation of the common carotids and in the aortic arch
What do the baroreceptors respond to?	Carotid sinus baroreceptors: any change in stretch Aortic arch receptors: increases in blood pressure
What nerves do the baroreceptors utilize to regulate blood pressure?	Carotid sinus baroreceptors: glossopharyngeal nerve (CN IX) Aortic arch baroreceptors: vagus nerve
What are the steps in the baroreceptor reflex?	Decreased stretch on the walls due to decreased arterial pressure → decreases firing rate of the afferent limb of the reflex Rate determines the autonomic response coordinated by the vasomotor center to maintain blood pressure at normal
What mediates the response of the vasomotor center?	Decreased parasympathetic stimulation and increased sympathetic stimulation
What autonomic responses does the vasomotor center utilize to maintain blood pressure?	1. Increased HR 2. Increased contractility 3. Increased SV 4. Vasoconstriction of arterioles and veins
What happens during carotid massage?	Massage increases pressure on carotid artery, which is interpreted as increased stretch and leads to decreased HR

Cardiovascular Physiology

Where are the peripheral chemoreceptors located?	Carotid and aortic bodies
What do they respond to?	1. Decreased P_{O_2} 2. Decreased pH of blood 3. Increased P_{CO_2}
Where are the central chemoreceptors located?	Vasomotor center
Do the central chemoreceptors respond directly to P_{O_2}?	No
What do they respond to?	Changes in pH and P_{CO_2} of brain interstitial fluid
What is the Cushing reaction?	Response to cerebral ischemia where increased intracranial pressure causes hypertension (sympathetic) and bradycardia (parasympathetic)
What is the body's long-term blood pressure regulation system?	Renin-angiotensin-aldosterone system
How does it regulate blood pressure?	By adjusting blood volume
Describe how each of following are involved in the renin-angiotensin-aldosterone system:	
Juxtaglomerular complex	Detects decreased renal perfusion in the afferent arteriole, causing them to release renin
Renin	Catalyzes the conversion of circulating angiotensinogen into angiotensin I
Angiotensin-converting enzyme (ACE)	Converts angiotensin I into angiotensin II
Where is ACE primarily located?	Lungs
What are the effects of angiotensin II?	1. Stimulates synthesis and secretion of aldosterone, which increases Na^+ reabsorption and K^+ secretion in the kidney 2. Stimulates vasoconstriction of the arterioles, which increases TPR and MAP
What is the other name for antidiuretic hormone (ADH)?	Vasopressin
Where is ADH secreted?	Posterior pituitary
What causes the release of ADH?	Atrial receptors detect decrease in blood volume or blood pressure (e.g., hemorrhage)

What are the effects of ADH?	1. Increases TPR by activating V_1 receptors in the arterioles causing them to vasoconstrict 2. Increases water reabsorption by the renal distal tubule and collecting ducts with the activation of V_2 receptors
What causes the release of atrial natriuretic peptide (ANP)?	Increased atrial pressure
Where is ANP released?	The atria
What are the effects of ANP?	1. Dilation of arterioles and decreased TPR from relaxation of vascular smooth muscle 2. Increased Na^+ and H_2O excretion by the kidney to decrease blood volume 3. Inhibits renin secretion

SPECIAL CIRCULATIONS

What is autoregulation?	Mechanism by which blood flow is altered to meet the demands of the tissue
What organs exhibit autoregulation?	Brain, heart, and kidney
What is active hyperemia?	Blood flow to the organ is proportional to the metabolic activity of the organ
What is reactive hyperemia?	Transient increase in blood flow to an organ after it has undergone a brief period of arterial occlusion (e.g., ischemia)
What mechanisms have been proposed to explain the local control of blood flow?	1. Myogenic hypothesis: based on vascular smooth muscle contracting when stretched and explains autoregulation, but not active or reactive hyperemia 2. Metabolic hypothesis: based on tissue supply of oxygen matching the tissue's demand and the production of vasodilator metabolites (e.g., CO_2, H^+, K^+, lactate, adenosine)

What effects do each of the following have on the vasculature:	
Histamine and bradykinin	Arteriolar dilation and venous constriction
5-Hydroxytryptamine (5-HT, serotonin)	Arteriolar constriction
Prostacyclin and E-series prostaglandins (PGE_1 and E_2)	Vasodilation
F-series prostaglandins and thromboxane A_2	Vasoconstriction
What factors determine autoregulation in the following locations:	
Brain	Local metabolic factors: P_{CO_2}
Heart	Local metabolic factors: hypoxia, adenosine, and nitrous oxide (NO)
Kidney	Myogenic and tubuloglomerular feedback
What factors determine autoregulation in skeletal muscle at rest and with exercise?	At rest: sympathetic innervation With exercise: local metabolic factors: lactate, adenosine, and K^+

MICROCIRCULATION AND LYMPHATICS

What is located at the junction between arterioles and capillaries?	Precapillary sphincter
Do capillaries contain smooth muscle?	No
What are capillaries composed of?	Single layer of endothelial cells and a surrounding basement membrane
How do small water-soluble substances enter capillaries?	Through pores (clefts) between adjacent endothelial cells
Can proteins fit through the pores?	Not normally, they are too large
Where are the pores the tightest?	In the brain, they form the blood-brain barrier
What are the components of the blood-brain barrier?	1. Endothelial cells of the cerebral capillaries 2. Choroid plexus epithelium
What substances pass readily through the barrier?	Lipid-soluble substances
What substances are excluded by the barrier?	Protein and cholesterol

What are the functions of the blood-brain barrier?	1. Maintains a constant environment for the neurons in the central nervous system (CNS) 2. Protects the brain from toxic substances 3. Prevents movement of neurotransmitter into the circulation

For the following compounds indicate where their concentration is higher, blood or cerebrospinal fluid (CSF):

Ca^{2+}	Blood
Cholesterol	Blood (absent in CSF)
Cl^-	Equal in both
Creatinine	CSF
Glucose	Blood
HCO_3^-	Equal in both
K^+	Blood
Mg^{2+}	CSF
Na^+	Equal in both
Protein	Blood (absent CSF)
Where are the pores the widest?	In the liver and intestine (sinusoids)
What is unique about the junctions in the sinusoids?	They are wide enough to allow the passage of proteins
How do large water-soluble substances enter capillaries?	Pinocytosis
How do lipid-soluble substances enter capillaries?	Through the membrane of the endothelial cells by simple diffusion
What equation governs fluid exchange across capillaries?	The Starling equation: $$J_v = K_f [(P_c - P_i) - (\pi_c - \pi_i)]$$ where J_v = fluid movement (mL/min), K_f = hydraulic conductance (mL/min · mm Hg), P_c = capillary hydrostatic pressure (mm Hg), P_i = interstitial hydrostatic pressure (mm Hg), π_c = capillary oncotic pressure (mm Hg), π_i = interstitial oncotic pressure (mm Hg)

What can cause an increase in P_c?	Increased arterial or venous pressure
What can cause a decrease in π_c?	Decreased blood protein concentration
What can cause an increase in π_i?	Inadequate lymphatic function
What is lymph?	The excess fluid that is filtered out of capillaries that is not reabsorbed
What is the function of lymph?	To return the lost fluid along with any filtered proteins back into circulation
What type of flow do the lymphatics demonstrate?	Unidirectional
What permits this type of flow?	One-way flap valves
What causes edema?	There is more interstitial fluid than the lymphatics can return into the circulation

CHAPTER 4
Respiratory Physiology

LUNG VOLUMES AND CAPACITIES

Define the following lung volumes:

Tidal volume (V_T) — Volume of a normal breath at rest (average 500 mL)

Inspiratory reserve volume (IRV) — Additional volume of gas that can be inspired above the V_T (average 3 L)

Expiratory reserve volume (ERV) — Volume of gas that can be forcefully expired after a normal passive expiration (average 1.3 L)

Residual volume (RV) — Volume of gas that remains after maximal expiration (average 1.5 L)

How are lung volumes and capacities related? — Capacity = sum of ≥ 2 volumes

Define the following lung capacities:

Total lung capacity (TLC) — Volume of gas present after a maximal inspiration (average 6 L)

Vital capacity (VC) — Maximal volume that can be expelled after a maximal inspiration (average 4.5 L)

Functional residual capacity (FRC) — Volume remaining at the end of a normal breath at rest (average 3 L)

What are the volumes that make up the following capacities:

Inspiratory capacity (IC) — $IC = V_T + IRV$

FRC — $FRC = RV + ERV$

VC — $VC = ERV + V_T + IRV$

TLC — $TLC = RV + ERV + V_T + IRV$

Identify the labeled lung volumes and capacities on the spirogram below:
1. V_T
2. IRV
3. ERV
4. Reserve volume
5. VC
6. FRC
7. TLC

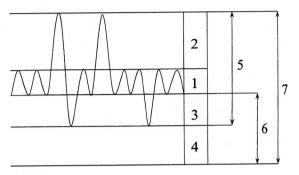

Figure 4.1 Spirogram.

Which lung volumes and capacities are estimated? — RV, FRC, and TLC

How does age affect the following parameters of pulmonary function tests (PFTs):

TLC	Decreased
RV	Increased
VC	Decreased
CV	Increased
FRC	Does not change

Define minute ventilation (V_E) — Volume of air inspired or expired per minute:

$$V_E = V_T \times \text{frequency}$$

What are:

Forced vital capacity (FVC) — Volume exhaled with maximal expiratory effort

Forced expiratory volume in 1 second (FEV_1)? — Volume that can be forcefully expired in 1 second

What is the normal percentage of FEV_1 per FVC?	75–80% (FEV_1/FVC = 0.8)
What is the main characteristic of obstructive lung disease?	Airflow limitation, usually not fully reversible and progressive and is associated with an abnormal inflammatory response to noxious stimuli (e.g., particles or gases)
What are the major characteristics of restrictive lung disease?	Inflammation and fibrosis of lung parenchyma resulting in abnormalities in gas exchange at rest or during exercise
What happens to FEV_1 and FVC in restrictive disease processes?	Both FEV_1 and FVC are reduced (FEV_1/FVC is normal or increased)
What happens to FEV_1 and FVC in obstructive disease processes?	FEV_1 is reduced more than FVC (FEV_1/FVC is decreased)
What type(s) of cells line the alveoli?	Two types of epithelial cells: Type I pneumocytes (90%): primary lining cells Type II pneumocytes (10%): granular cells
What is respiratory dead space?	Space in the conducting zone of the airways where no gas exchange occurs
What type(s) of dead space exists?	1. Anatomic dead space (e.g., conducting airways) 2. Alveolar dead space (e.g., alveoli with no blood flow)
How can you approximate the volume of anatomic dead space?	Dead space in milliliter ≈ body weight in pounds
What are the functions of anatomic dead space?	1. Conditions air (inspired gas is warmed and saturated with water vapor) 2. Removal of foreign materials
What is alveolar ventilation?	The amount of air reaching the alveoli per minute
How does dead space affect alveolar ventilation?	Some inspired gas is trapped in the dead space making alveolar ventilation less than the respiratory minute volume
How does shallow versus deep breathing affect alveolar ventilation?	Rapid shallow breathing produces much less alveolar ventilation

How can dead space (V_D) be calculated?	$P_ECO_2 \cdot V_T = P_aCO_2(V_T - V_D) + P_ICO_2 \cdot V_D$ where $P_ECO_2 = PCO_2$ of expired air, $P_aCO_2 = PCO_2$ of arterial blood. **Note:** P_ICO_2 (inspired PCO_2) $\times V_D$ is very small and can be ignored

RESPIRATORY MECHANICS

What characteristics of the lungs are involved in breathing?	1. Elastic work 2. Viscous resistance 3. Airway resistance
What is elastic work?	Stretching of elastic tissue in the chest wall and lungs by the respiratory muscles
What components in the lungs are major determinants of elastic forces of the lung tissue?	Elastin and collagen fibers interwoven among the lung parenchyma
What is viscous resistance?	Moving inelastic tissues
What percentage of O_2 consumption is required with normal breathing?	2–3%
What type of process is inspiration?	Active
Which muscles are involved in inspiration?	1. Diaphragm 2. External intercostals muscles 3. Accessory muscles
Which muscle contributes most to the inspiratory effort?	Movement of the diaphragm contributes ~75% of the change in intrathoracic volume during normal inspiration
Which nerve supplies the diaphragm?	The phrenic nerves: one-third of their fibers are sensory to the diaphragm and two-third of their fibers are motor (C3, C4, C5 keep the diaphragm alive)
What happens when the diaphragm contracts?	Volume of the thoracic cavity is increased as abdominal contents are pushed down and ribs are lifted upward and out
When are the other muscles of inspiration mainly used?	Exercise

What are the accessory muscles of respiration?	1. Sternocleidomastoid (SCM) 2. Scalenes 3. Strap muscles of the neck
Transection of the spinal cord at which level results in respiratory arrest?	Above the third cervical segment
What type of process is expiration?	Passive
Which muscles are involved in active expiration?	Internal intercostals muscles and anterior abdominal wall muscles
When does active expiration occur?	Exercise and when airway resistance is increased
Define compliance (C).	An indication of how easily the lungs and chest wall can be stretched or inflated
What is the equation for compliance?	$C = \Delta V / \Delta P$ where ΔV = change in volume, ΔP = change in pressure
What affects compliance?	Lung volume and alveolar surface tension
What processes can cause a decrease in compliance?	Pulmonary congestion and interstitial pulmonary fibrosis
How does emphysema affect compliance?	Increases it
What is intrapleural pressure?	Pressure of the fluid in the space between the lung pleura and the chest wall pleura
What is the intrapleural pressure at the base of the lungs?	−2.5 mm Hg relative to atmospheric pressure
What contributes to the intrapleural subatmospheric pressure?	The lungs' passive recoil properties and the stiff chest wall's tendency to expand at physiologic chest and lung volumes
Define elasticity.	The recoil force generated by distension of a structure
What is the relationship between elastance and compliance?	Inverse: $C = 1/E$
What contributes to the lungs' recoil properties?	Elastic tissue and surface tension
How do Laplace's law and surface tension affect the collapsibility of alveoli?	$P = \dfrac{2T}{r}$ where P = collapsing pressure (dyn/cm^2), T = surface tension (dynes/cm), and r = alveolar radius (cm)

Which is easier to keep open, a large alveoli or a small one?	Large alveoli (alveolar radius is inversely proportional to collapsing pressure)
What is surfactant composed of?	1. Dipalmityl phosphatidylcholine (aka lecithin)—major component 2. Phosphatidylglycerol 3. Other lipids 4. Neutral lipids 5. Proteins
What are the functions of pulmonary surfactant?	1. Reduce surface tension at low lung volumes (prevent atelectasis) 2. Increase surface tension at high lung volumes (contribute to lung recoil) 3. Increase alveolar radius 4. Reduce pulmonary capillary infiltration
Which cells produce surfactant?	Type II alveolar epithelial cells
How may surfactant synthesis be reduced?	1. Developmental deficiency (e.g., prematurity) 2. Hypovolemia 3. Hypothermia 4. Acidosis 5. Hypoxemia 6. Rare genetic disorders of surfactant synthesis
What is the significance of surfactant in infant respiratory distress syndrome (IRDS)?	There is a surfactant deficiency, which results in high surface tension in the lungs, as well as many areas of atelectasis. There is also decreased FRC and subsequent arterial hypoxemia
What does the therapy for IRDS include?	1. Positive end-expiratory pressure (PEEP) 2. Exogenous surfactant 3. Steroids
By what week do the fetal lungs make surfactant?	Week 34–36
What may indicate fetal pulmonary maturity?	The ratio of lecithin concentration (which increases with maturity) to sphingomyelin (which remains constant during gestation) or L/S ratio in the amniotic fluid. The presence of minor phospholipids (e.g., phosphatidylglycerol) is also indicative in cases where the L/S ratio is borderline

What L/S ratio usually indicates pulmonary maturity?	2:1
What can be used to help accelerate the maturation of surfactant in the lungs of a fetus?	Glucocorticoid hormones
How are individual alveoli prevented from collapsing normally?	Alveolar walls and airway walls are structurally connected so that tension on adjacent walls created by collapsing an alveolus holds it open
What factors determine airway resistance?	1. Rate of gas flow through the airway 2. Diameter of the airway 3. Length of the airway
What law describes airway resistance?	Poiseuille's law: $$R = \frac{8\eta l}{\pi r^4}$$ where R = resistance, η = viscosity of inspired gas, l = length of airway, and r = radius of airway
How are airway resistance and airflow related?	Inversely: $$Q = \frac{\Delta P}{R}$$ where Q = airflow (L/min), ΔP = pressure gradient (cm H_2O), and R = airway resistance (cm/H_2O/L/min)
Which part of the respiratory system is the major site of airway resistance?	Medium-sized bronchi
Which part of the respiratory system has the highest individual resistance?	Small terminal airways (they are not the major site of airway resistance due to their parallel arrangement)
What factors can change airway resistance?	1. Altering the radius of the airways 2. Changes in lung volume 3. Viscosity/density of the inspired gas
What causes bronchoconstriction?	1. Parasympathetic discharge 2. Substance P 3. Adenosine 4. Hypersensitivity response (e.g., histamines) 5. Arachidonic acid metabolites (e.g., prostaglandins and leukotrienes)

How does bronchoconstriction affect airways?	1. Reduces airway radius 2. Increases resistance 3. Limits airflow during inspiration or expiration
What causes bronchodilation?	Sympathetic discharge and sympathetic agonists via β_2 receptors
How do obstructive diseases affect respiratory mechanics?	Increase airway resistance
How do restrictive diseases affect respiratory mechanics?	Decrease compliance

GAS EXCHANGE AND TRANSPORT

How are partial pressures determined?	Dalton's law: Partial press = Total pressure × fractional [gas] where press = pressure and fractional [gas] = gas concentration
What are the partial pressures of O_2 and CO_2 in the following locations:	
Atmospheric air	O_2: 160 mm Hg, CO_2: 0 mm Hg
Air in the trachea	O_2: 150 mm Hg, CO_2: 0 mm Hg
Alveolar air	O_2: 100 mm Hg, CO_2: 40 mm Hg
Arterial blood	O_2: slightly <100 mm Hg, CO_2: 40 mm Hg
Mixed venous blood	O_2: 40 mm Hg, CO_2: 46 mm Hg
Why is the P_{O_2} in the trachea less than that of the atmosphere?	The air in the trachea is humidified (addition of H_2O, which decreases P_{O_2})
Why is the P_{O_2} in arterial blood slightly less than 100 mm Hg?	*Physiologic shunt*
What is a *physiologic shunt*?	The ~2% of systemic cardiac output that bypasses the pulmonary circulation
True or false? The composition of alveolar gas remains relatively constant at rest.	True. The average FRC is 2.8 L, thus, each increment of inspired and expired air (average 500 mL) has relatively little effect on P_{O_2} and P_{CO_2}

Define gas exchange. — Transport of gas from alveoli to hemoglobin (Hb) in the blood across the respiratory membrane

Where in the respiratory system does gas exchange occur? — In the terminal portions of the airways (respiratory bronchioles, alveolar ducts, and alveoli)

At rest, how long does it take for the blood to traverse the pulmonary capillaries? — 0.75 second

What factors determine pulmonary gas diffusion? — Partial pressure differences across the membrane and the area available for diffusion

What equation governs pulmonary gas diffusion? — Fick's equation:

$$V_{gas} \propto \left(\frac{A}{Th}\right) \times D \times (P_C - P_A)$$

where V_{gas} = flow of gas across an area, A = area of barrier, Th = thickness of barrier, D = diffusion coefficient, P_C = partial pressure of gas in the pulmonary capillary, P_A = partial pressure of gas in the alveoli

Define the diffusion coefficient. —

$$D \propto \frac{Sol_{gas}}{\sqrt{MWt}}$$

where D = diffusivity factor, Sol_{gas} = gas solubility in tissue fluid, and MWt = molecular weight

Explain flow limitation. — Substances crossing the alveolar membrane do not react to substances in the blood and reach equilibrium in less time than it takes for blood to traverse the pulmonary capillaries

What is another name for flow limitation? — Perfusion limitation

How can flow limitation be overcome? — By increasing blood flow

Explain diffusion limitation. — Substances react with substances in the blood but equilibrium is not reached in the time it takes for blood to traverse the pulmonary capillaries

What is the significance of diffusing capacity (D_L)?	Its measurement permits evaluation of the diffusion properties of alveolar-capillary membrane by measuring the rate of gas transfer (conductance) by the respiratory system
What factors affect D_L?	The diffusion properties of the system (e.g., membrane component) and the rate of the chemical reaction of the system (e.g., blood component)
How is O_2 transported in blood?	In arterial blood at P_{O_2} 95 mm Hg, P_{CO_2} 40 mm Hg, and Hb 97% saturated
	Major: Chemical combination with Hb (19.5 mL O_2/100 mL blood)
	Minor: Dissolved in the plasma (0.29 mL O_2/100 mL blood)
What is the normal O_2 tension of arterial blood?	85–100 mm Hg
In normal adults, what are hemoglobin molecules composed of?	Two α and two β chains
What is heme?	Complex made up of a porphyrin ring and one atom of ferrous iron
What is the role of the ferrous iron (Fe^{2+})?	Each ferrous iron binds one O_2 molecule reversibly
What is the significance of the iron in heme?	It stays in the ferrous state, so that binding to O_2 does not result in its oxidation, but rather in its oxygenation
How does Hb affect the O_2 content in blood?	Increases the amount of O_2 that is carried in the blood compared to plasma almost 70-fold
What percent of O_2 in arterial blood is carried by hemoglobin?	~98.5%
Define hemoglobin saturation.	Percent of hemoglobin that is combined with O_2
What influences the amount of O_2 that combines with hemoglobin?	O_2 tension (P_{O_2}) or O_2 saturation
What is the upper limit for O_2 tension in the body?	100 mm Hg

Define O₂ capacity.	The maximal amount of O_2 that can be carried in the blood by hemoglobin
What is the hemoglobin-oxygen (Hb-O) dissociation curve?	The curve relating percentage saturation of the O_2-carrying capacity of hemoglobin to the P_{O_2}

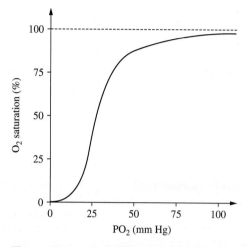

Figure 4.2 Hb-O Dissociation Curve.

How does the hemoglobin structure affect its oxygen capacity?	Hemoglobin has a quaternary structure that varies in its affinity for oxygen depending on the number of oxygen bound to it
How does the Hb-O dissociation curve affect the hemoglobin subunits in deoxygenated state?	Hemoglobin subunits are in a *tense* (*tight*) configuration where they have a reduced affinity for O_2
How does the Hb-O dissociation curve affect the hemoglobin subunits when oxygen is bound to it?	With each progressive O_2 molecule, hemoglobin takes on a *relaxed* configuration where subsequent binding of O_2 is facilitated as each O_2-binding site is exposed (called allosterism)
How does the sigmoid curve affect oxygen delivery at the:	
Plateau	High P_{O_2} gives a large reserve during oxygen loading
Steep section	Small change in P_{O_2} results in large changes in O_2 saturation. This facilitates unloading of oxygen at tissues

Where are the following parts of the sigmoid curve found in the human body?	
Plateau	Pulmonary capillaries
Steep section	Tissue capillaries
What does decreased oxygen affinity mean?	A higher P_{O_2} is required for hemoglobin to bind a given amount of O_2
What factors decrease the affinity of hemoglobin for oxygen?	1. Low pH (acidosis—increased H^+) 2. Increased P_{CO_2} 3. Increased temperature 4. Increased 2, 3-diphosphoglycerate (2, 3-DPG) concentration
What factors cause an increase in 2, 3-DPG concentrations?	1. High pH 2. Thyroid hormone 3. Growth hormone 4. Androgens 5. High altitudes
What are the functions of hemoglobin?	1. Facilitates O_2 transport 2. Facilitates CO_2 transport 3. Acts as a buffer in the blood 4. Facilitates NO transport
What happens to the Hb-O dissociation curve with decreased affinity for oxygen by hemoglobin?	Curve shifts to the right
Diagram the changes in the Hb-O dissociation curve with decreased affinity for oxygen.	

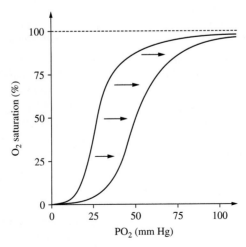

Figure 4.3 Hb-O with Shifts.

What does decreased oxygen affinity favor?	Oxygen delivery to tissue
What is the Bohr effect?	Decrease in affinity of Hb for oxygen due to lowered pH. This is related to the fact that deoxygenated Hb binds H^+ more avidly than does oxygenated Hb
Where is the Bohr effect used physiologically?	In the peripheral tissue (helps unload oxygen)
How does fetal hemoglobin (HbF) differ from adult hemoglobin (HbA)?	HbF contains γ polypeptide chains instead of β chains
What is the significance of the γ chains versus the β chains HbF?	There is poor binding of 2,3-DPG by the γ chains causing HbF to have greater affinity for O_2 than HbA
How does anemia affect 2,3-DPG concentration in red blood cells (RBCs)?	Increases it
How is myoglobin different from hemoglobin?	Myoglobin binds one rather than four molecules of O_2 per molecule, resulting in a rectangular hyperbolar dissociation curve, typical of Michaelis-Menten kinetics
How does the dissociation curve of myoglobin affects its O_2 uptake relative to hemoglobin?	Because its curve is to the left of the hemoglobin curve, it takes up O_2 from hemoglobin in the blood and releases O_2 only at low PO_2 values
Where can much of the myoglobin be found?	In skeletal muscles
What does decreased arterial O_2 content cause?	Decreased hemoglobin concentration and reduced arterial O_2 tension
Which does Hb have more affinity for CO or O_2?	CO (~200 × greater affinity)
What is formed as a result of CO reacting with hemoglobin?	Carboxyhemoglobin
How does carboxyhemoglobin formation affect O_2 content?	1. Decreases the functional Hb concentration 2. Reduces oxygen-carrying capacity of blood 3. Lowers the tissue O_2 tension
How is CO_2 transported in blood?	1. H^+ buffered as HCO_3^- in plasma (major) 2. Dissolved in plasma or in RBCs (minor) 3. Formation of carbamino-Hb in RBCs (minor)

	4. Formation of carbamino compounds with plasma protein (minor)
Where does hydration of CO_2 into HCO_3^- occur?	RBCs
What enzyme catalyzes the conversion of CO_2 into HCO_3^- ?	Carbonic anhydrase catalyzes formation of H_2CO_3 from H_2O and CO_2, which dissociates into H^+ and HCO_3^-
How does HCO_3^- enter the plasma?	Exchanged with Cl^- (chloride shift)
Which two factors determine arterial or alveolar CO_2 tensions.	Rate of CO_2 production and alveolar ventilation
What is the Haldane effect?	Binding of O_2 to hemoglobin reduces its affinity for CO_2
Where is the Haldane effect used physiologically?	Lungs
What is methemoglobin?	Hb that contains ferric (Fe^{3+}) iron. It has a very high affinity for oxygen, resulting in decreased ability of Hb to unload oxygen.
How do obstructive diseases affect gas exchange?	Increases the time constant and result in a slower rate of acinar filling and emptying
How do restrictive diseases affect gas exchange?	Reduce the time constant

PULMONARY CIRCULATION

Compare pulmonary versus systemic circulation.	Pressures and resistance are much lower in the pulmonary circulation	
Compare the following parameters in adult and fetal pulmonary circulation:	Adult	Fetal
Pressure	Low	High
Resistance	Low	High
Flow	High	Low
What happens to resistance as pressure in the pulmonary artery increases?	Resistance decreases even lower because previously collapsed capillaries open up (recruitment) and individual capillary segments widen (distension)	

Respiratory Physiology

Where does pulmonary blood flow come from?	Cardiac output of the right ventricle
What factors contribute to the large compliance of the pulmonary arterial tree?	1. Short pulmonary arterial branches 2. All pulmonary arteries have larger diameters than their counterpart systemic arteries 3. Vessels are very thin and distensible
How does blood flowing through the pulmonary arteries differ from that flowing through the systemic arteries?	Blood in the pulmonary arteries is partially deoxygenated, while that in the systemic arteries is oxygenated
True or false? Blood flowing through the bronchial arteries is partially deoxygenated.	False
What does bronchial arterial blood supply?	Supporting tissues of the lungs including the connective tissue, septa, and large and small bronchi
True or false? All of the blood in the left ventricle comes from the right ventricle.	False. Some of the blood enters the bronchial blood flow and blood that flows through the coronary arteries enters the left side of the heart without entering the right atrium
At which point is the pressure in the pulmonary artery essentially equal to the pressure in the right ventricle?	During the systolic phase of the cardiac cycle
What are the average pulmonary arterial pressures?	Systole: 25 mm Hg Diastole: 8 mm Hg
How is the pulmonary wedge pressure obtained?	Via a catheter introduced via the systemic venous circulation (e.g., femoral vein) that is *wedged* in a pulmonary arteriole
What does the pulmonary wedge pressure estimate?	1. Pulmonary venous pressure 2. Left atrial pressure 3. Left ventricular end-diastolic pressure
How does gravity affect the pulmonary circulation?	In the upright position, the apices of the lungs are above the level of the heart, while the bases are at or below it. Consequently, there is a relatively marked pressure gradient in the pulmonary arteries
When standing, where is pulmonary blood flow the least?	Apices

What happens when pressure in the pulmonary artery increases?	Resistance falls even lower because of recruitment of previously collapsed capillaries and distension of existing capillary segments
What are some consequences of long-standing pulmonary hypertension?	Cor pulmonale and right ventricular failure
What is the physiologic response of the lungs to hypoxia?	Vasoconstriction
Why is the response to hypoxia by the lungs significant?	It redirects blood from poorly ventilated regions to well-ventilated areas
What changes does chronic hypoxia cause in the pulmonary circulation?	1. Sustained pulmonary vasoconstriction 2. Increase in number of perivascular smooth muscle cells 3. Increased interstitial collagen
How does inspiration affect venous return?	Increases it by increasing intrapleural pressure

VENTILATION AND PERFUSION

Which structures in the lungs are involved in gas transport?	Conducting airways (including trachea, bronchi, bronchioles, and terminal bronchioles)
What is the major function of the lungs?	To provide oxygen for tissue metabolism via ventilation and gas exchange
Define ventilation (V).	Transport of gas from the environment to the alveoli for gas exchange
Define perfusion (Q).	Pulmonary blood flow to the alveoli
What is the most common cause of hypoxemia?	V/Q mismatch
What are some major causes of V/Q mismatch?	1. Fibrosing mediastinitis 2. Effusion 3. Vasculitis 4. One pulmonary artery (e.g., hyperplasia) 5. Neoplasm 6. Embolism (Remember the mnemonic: **FEV ONE**)

How do ventilation and perfusion in the upright position change from the bases to the apices of the lungs?	Decline
What is the normal distribution of ventilation and perfusion in a standing person?	Ventilation: highest at base, lowest at apex Perfusion: highest at base, lowest at apex
Why is the V/Q ratio low at the base and high at the apex?	The relative change in blood flow from the apex to the base is greater than the relative change in ventilation
What happens to P_{O_2} and P_{CO_2} if the ventilation to an alveolus is reduced relative to its perfusion?	P_{O_2}: falls because less O_2 is delivered P_{CO_2}: rises because less CO_2 is expired
What is a shunt?	Perfusion with no/low ventilation
What is the Alveolar-arterial (A-a) gradient?	The discrepancy between alveolar and arterial oxygen partial pressures
How is the A-a gradient calculated?	$P_AO_2 - P_aO_2 = P_IO_2 - \left(\frac{P_aCO_2}{R}\right) - P_aO_2$ where P_AO_2 = alveolar P_{O_2}, P_aO_2 = arterial P_{O_2}, $P_IO_2 = P_{O_2}$ of inspired gas, P_aCO_2 = arterial P_{CO_2}, and R = gas constant
At what V/Q ratio does the most efficient gas exchange occur?	1
What is the V/Q ratio for the whole lung at rest?	Ventilation = 4.2 L/min Perfusion = 5.5 L/min V/Q = 4.2/5.5 ≈ 0.8
What causes low V/Q ratio?	Inadequate ventilation or excessive blood flow
What is the result of low V/Q ratio?	Decreased alveolar O_2 tension and increased CO_2 tension

RESPIRATORY CONTROL

Where is respiration centrally controlled?	Reticular formation in the medulla
Where do the nerve fibers that mediate inspiration converge?	On the phrenic motor neurons located in the ventral horns from C3 to C5 and the external intercostals motor neurons in the ventral horns throughout the thoracic cord

Where do fibers concerned with expiration converge?	Primarily on the internal intercostal motor neurons in the thoracic cord
What is the dorsal respiratory group (DRG)?	Inspiratory cells that may act as the primary rhythm generator for respiration
What stimulates DRG activity?	1. Low O_2 tensions 2. High CO_2 tensions 3. Low pH levels 4. Increased electrical traffic in renal artery stenosis (RAS)
What nerves mediate input to the DRG?	Vagus: peripheral chemoreceptors and lung mechanoreceptors Glossopharyngeal: peripheral chemoreceptors
Where does the outflow from DRG project?	Contralateral phrenic and intercostals motor neurons, and the ventral respiratory group (VRG)
What makes up the VRG?	Upper motor neurons of the vagus and the nerves to the accessory muscles of respiration
What is the role of the VRG?	Activated to control expiration when it is an active process (e.g., exercise)
Where is the apneustic center?	Caudal area of the lower pons
What is the significance of the apneustic center?	Efferent outflow increases the duration of inspiration
Where is the pneumotaxic center?	Upper part of the pons
What is the function of the pneumotaxic center?	Unknown, it is thought to inhibit the apneustic center and shortens inspiration. It may play a role in switching between inspiration and expiration
How does damage to the pneumotaxic center affect respiration?	Respiration becomes slower and V_T greater
Where are central chemoreceptors located?	Beneath the ventral surface of the medulla
What do central chemoreceptors respond to?	H^+ concentration in the cerebrospinal fluid (CSF) and the surrounding interstitial fluid
What is the major chemical drive of respiration?	CO_2 (H^+) effects on the central chemoreceptors
Where are the peripheral chemoreceptors located?	Carotid and aortic bodies

Respiratory Physiology

What do peripheral chemoreceptors respond to?	1. Lowered O_2 tensions 2. Increased CO_2 tensions 3. Increased H^+ concentrations in arterial blood
What stimuli affect the respiratory center?	Chemical control: 1. CO_2 (via CSF and brain interstitial fluid H^+ concentration) 2. O_2 (via carotid and aortic bodies) 3. H^+ (via carotid and aortic bodies) Nonchemical control: 1. Vagal afferents from receptors in the airways and lungs 2. Afferents from the pons, hypothalamus, and limbic system 3. Afferents from proprioceptors 4. Afferents from baroreceptors
Define apnea.	Cessation of respiration lasting >20 seconds
What is obstructive sleep apnea (OSA)?	Recurrent interruptions of breathing during sleep due to temporary obstruction of the airway by lax, excessively bulky, or malformed pharyngeal tissues (soft palate, uvula, and sometimes tonsils), with resultant hypoxemia and chronic lethargy
How does OSA differ from central sleep apnea?	OSA: obstruction with respiratory effort (e.g., chest movement) Central sleep apnea: apnea without respiratory effort

RESPONSE TO STRESS

How does the respiratory system respond to exercise?	1. Increase minute ventilation 2. Increase CO_2 output 3. Increase O_2 consumption 4. A-a gradient widens (excessive exercise) 5. Respiratory (CO_2/O_2) exchange ratio exceeds 1, but <1.25

How does the amount of O_2 in the body change with exercise?	Increases because the amount of O_2 added to each unit of blood and the pulmonary blood flow per minute are increased
How does the P_{O_2} of blood flowing into the pulmonary capillaries change with exercise?	Falls to 25 mm Hg or less because of increased extraction
How does CO_2 excretion change with exercise?	Increases to as much as 40-fold because of increased amount of CO_2 produced
What happens to the mean values for arterial P_{O_2} and P_{CO_2} during exercise?	No change
What happens to the level of lactate in the blood with exercise?	It increases
Where does lactate come from?	Muscles in which aerobic resynthesis of energy stores cannot keep pace with their utilization and an oxygen debt is incurred
What happens to arterial pH during exercise?	No change with moderate exercise, but it decreases with strenuous exercise due to lactic acidosis
What is hypoxia?	O_2 deficiency at the tissue level
What are the signs of hypoxia?	Cyanosis, tachycardia, and tachypnea
What are the symptoms of chronic hypoxia?	Dyspnea and shortness of breath
Define dyspnea.	Difficult or labored breathing in which the subject is conscious of shortness of breath
What causes cyanosis?	Reduced hemoglobin, which has a dark color and causes a dusky bluish discoloration of the tissues
What physiologic changes occur at high altitudes?	Alveolar P_{O_2} decreases from decreased barometric pressure, which results in decreased arterial P_{O_2} (e.g., hypoxia)
What is the acute response to high altitudes?	Increased ventilation rate (e.g., hyperventilation)

What are the chronic responses to high altitudes?	1. Increased ventilation rate 2. Increased erythropoietin production by kidneys, which increases Hb concentration. 3. Increased 2, 3-DPG 4. Increased number of mitochondria in cells 5. Increased renal excretion of HCO_3^-
What agent can be given to treat the respiratory alkalosis caused by the hyperventilation of hypoxia?	Acetazolamide, which inhibits carbonic anhydrase resulting in increased urinary excretion of HCO_3 and increased P_{ACO_2}
What is a consequence of chronic hypoxic pulmonary vasoconstriction?	Right ventricular hypertrophy

CHAPTER 5
Renal and Acid-Base Physiology

BODY FLUIDS

How can you approximate total body water (TBW)?	TBW is ~60% of the body weight
What is the distribution of water in the human body?	Intracellular fluid (ICF) is two-thirds of TBW Extracellular fluid (ECF) is one-third of TBW • Plasma is one-fourth of ECF or one-twelfth of TBW • Interstitial fluid is three-fourths of ECF or one-fourth of TBW
What are the major cations of the ICF and ECF?	ICF: K^+ and Mg^{2+} ECF: Na^+
What are the major anions of the ICF and ECF?	ICF: protein and organophosphates ECF: organophosphates and very little protein
What substance is used to measure the following major fluid compartments:	
TBW	Tritiated H_2O or D_2O
ECF	Sulfate, inulin, or mannitol
Plasma	Radioiodinated serum albumin, Evans blue
Interstitial fluid (IF)	*Indirectly*: IF = ECF − plasma
ICF	*Indirectly*: ICF = TBW − ECF

RENAL FILTRATION AND BLOOD FLOW

What is the equation used to measure clearance (C) of the kidney?

$$C = \frac{U \times V}{P}$$

where C = clearance, U = urine concentration, V = urine flow rate, and P = plasma concentration

If the urine concentration of inulin is 0.25 mg/mL, the urine flow rate is 25 mL/min and the plasma concentration of the drug is 0.12 mg/mL, what is the renal clearance?

Using

$$C = \frac{U \times V}{P}$$

and plugging in U = 0.25 mg/mL, V = 25 mL/min, and P = 0.12 mg/mL, we get [(0.25 mg/mL) × (25 mL/min)]/(0.12 mg/mL) = 52 mL/min

What percentage of the cardiac output goes to the kidneys?

25%

What factors influence renal blood flow (RBF)?

1. Difference between renal artery and renal vein pressures (direct effect)
2. Renal vasculature resistance (inverse relationship)

What is autoregulation of RBF?

Process by which renal vasculature changes pressure to keep the RBF constant

Describe the mechanisms by which autoregulation is accomplished.

Myogenic mechanism: stretch receptors in renal arterioles are activated (i.e., stretched from increased pressure) and cause the arterioles to contract, to increase resistance, and maintain constant blood flow

Tubuloglomerular feedback: when the renal arterial pressures increase, there is an increase in the fluid load. This increase is sensed by the macula densa, which causes the constriction of the nearby afferent arterioles to increase resistance and maintain constant blood flow

What substance is used to measure renal plasma flow (RPF)? Why?

Paraaminohippuric acid (PAH), it is filtered and secreted by renal tubules

How is RPF calculated?	Using the clearance equation with PAH $$RPF = C_{PAH} = \frac{[U]_{PAH} \times V}{[P]_{PAH}}$$ where C_{PAH} = clearance of PAH, $[U]_{PAH}$ = urine concentration of PAH, V = urine flow rate, and $[P]_{PAH}$ = plasma concentration of PAH
How do you measure RBF?	$$RBF = \frac{RPF}{(1 - \text{hematocrit})}$$ where (1 − hematocrit) is the fraction of blood occupied by plasma
What substance is used to measure the glomerular filtration rate (GFR)? Why?	Inulin, it is filtered but not reabsorbed or secreted by the renal tubules
What equation is used to measure GFR?	Using the clearance equation with inulin $$GFR = C_{inulin} = \frac{[U]_{inulin} \times V}{[P]_{inulin}}$$ where C_{inulin} = clearance of inulin, $[U]_{inulin}$ = urine concentration of inulin, V = urine flow rate, and $[P]_{inulin}$ = plasma concentration of inulin
What can be used clinically to approximate GFR?	Creatinine (Cr) clearance
Does GFR remain constant with aging?	No, it decreases with age
What happens to blood urea nitrogen (BUN) and plasma Cr when GFR decreases?	Increases, since both are filtered by the glomerulus
What happens to plasma Cr with age?	It remains relatively constant despite the decrease in GFR with age due to the decrease in muscle mass seen with aging
What are Starling forces?	The collection of factors that are the driving force for glomerular filtration
What is the Starling equation?	$GFR = K_f [(P_{GC} - P_{BS}) - (\pi_{GC} - \pi_{BS})]$

Describe each of the following terms from the Starling equation:

K_f — Filtration coefficient across the glomerular barrier

P_{GC} — Hydrostatic pressure in the glomerular capillary. This value is constant along the length of the capillary

P_{BS} — Hydrostatic pressure in Bowman's space

π_{GC} — Oncotic pressure in the glomerular capillary. It increases along the length of the glomerular capillary because water is filtered off, which increases the concentration of protein and increases the oncotic pressure

π_{BS} — Oncotic pressure in Bowman's space

What are the components of the glomerular filtration barrier?
1. Fenestrated capillary endothelium
2. Fused basement membrane with heparan sulfate
3. Epithelial layer consisting of podocyte foot processes

What is the normal value of π_{BS}? — Zero. The protein concentration in Bowman's space is so small that it may be ignored

What can cause an increase in P_{GC}? What is the effect on GFR? — Dilation of the afferent arteriole or constriction of the efferent arteriole. This will increase GFR

What can cause an increase in P_{BS}? What is the effect on GFR? — Constriction or blockage of the ureters. This will increase GFR

What can cause an increase in π_{GC}? What is the effect on GFR? — Increase in protein concentration. This will decrease GFR

What aspect of the filtration barrier restricts plasma proteins from being filtered? — Anionic glycoproteins (heparan sulfate) that line the barrier repel the negatively charged proteins

What is the filtration fraction (FF)? — The fraction of RPF that is filtered across the glomerular capillary

How is FF calculated? — $$FF = \frac{GFR}{RPF}$$

What is the normal value for FF? — 0.20 or 20%

What happens to the protein concentration in the peritubular capillaries with increasing FF? — Increases

Renal and Acid-Base Physiology

The following values were found while studying a glomerulus: afferent arteriole pressure = 30 mm Hg, oncotic pressure of glomerular capillary = 25 mm Hg, Bowman's space pressure = 12 mm Hg. From these values, determine whether filtration will be favored or not.

Net pressure is determined from the Starling equation by excluding the filtration constant, and recognizing that π_{BS} is zero.
The equation can be written as:
net pressure = $(P_{GC} - P_{BS}) - (\pi_{GC} - 0)$
Plugging the numbers in
$(30 - 12) - (25) = -7$ mm Hg
Therefore, filtration is not favored. A positive number represents a net pressure that favors filtration, because the factors promoting filtration outweigh those that inhibit it. In this case they do not

RENAL SECRETION, REABSORPTION, AND EXCRETION

How is the rate at which a molecule is filtered across the glomerular capillary calculated?

Rate filtered = GFR × $[\text{plasma}]_{\text{substance}}$
where $[\text{plasma}]_{\text{substance}}$ = plasma concentration of the substance

How is the rate of a molecule's excretion in urine calculated?

Rate excreted = V × $[\text{urine}]_{\text{substance}}$
where V = urine flow rate, $[\text{urine}]_{\text{substance}}$ = urine concentration of the substance

How do you determine if a substance is ultimately secreted or absorbed?

Take the difference between rate filtered and rate excreted:
If > 0 → secretion
If < 0 → absorption
If = 0 → neither secreted nor absorbed

Is glucose normally secreted or reabsorbed?

Reabsorbed

How?

There are Na^+-glucose cotransporters in the proximal tubules, though there are only a limited number of them

At what concentration of glucose do the Na^+-glucose carriers in the proximal tubules start to become saturated? What happens to any glucose above this level?

250 mg/dL
Above this level, glucose is excreted in the urine (e.g., glucosuria). At 350 mg/dL, the carriers become completely saturated and all filtered glucose is excreted in the urine

What is the term for the saturation point?	Transport maximum (T_m)
What is the term for the point below which all of the substance is reabsorbed?	Threshold

Draw the titration curve for glucose.

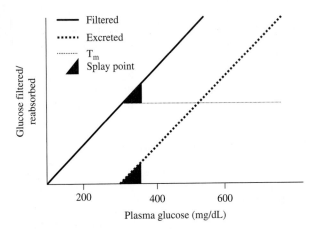

Figure 5.1 Glucose Titration.

What is splay?	The region on a titration curve between the threshold and T_m. In this region some of the substance is excreted despite not fully saturating reabsorption
What is the splay for glucose?	Between 250–350 mg/dL
What happens to the filtered load of PAH as the plasma load of PAH increases?	Increases in direct proportion to the plasma load
Is PAH secreted or reabsorbed?	Secreted
Where does this occur?	Proximal tubules
What is the mechanism by which PAH is secreted?	It is secreted into the tubular fluid (TF) (e.g., urine) by carriers in the proximal tubules
What happens to PAH secretion at low plasma concentrations?	Secretion rate will be low, but it will increase as plasma levels increase

Renal and Acid-Base Physiology

What happens to secretion of PAH when T_m is reached?	It plateaus and any further increase in plasma PAH will not change the secretion rate
How is excretion calculated for PAH (and other secreted substances)?	It is the sum of the amount filtered across the glomerular capillaries plus that secreted by the carriers in the peritubular capillaries into the TF
What is the relationship of RPF measurement to T_m for PAH?	RPF can only be measured at plasma concentrations $< T_m$
Draw the titration curve for PAH.	

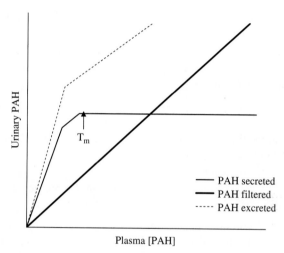

Figure 5.2 PAH Titration.

Will secreted substances have higher or lower clearance rates than the GFR?	Higher
Will reabsorbed substances have higher or lower clearance rates than the GFR?	Lower
What are some examples of substances with very low clearances?	1. Na^+ 2. Glucose 3. Amino acids 4. HCO_3^- 5. Cl^-
What does the ratio of TF to plasma (P) represent (TF/P)?	It compares the concentration of a substance in TF with that of plasma

What does each of the following ratios represent:

TF/P = 1 — Either there is no reabsorption of the substance or that reabsorption of the substance is equal to that of water

TF/P < 1 — Reabsorption of a substance is greater than water, thus concentration of TF is less than plasma

TF/P > 1 — Reabsorption of a substance is less than water, thus concentration of TF is greater than plasma

What substance is used as a marker for water reabsorption along the nephron? — Inulin

What is the equation to calculate the fraction of filtered water that has been reabsorbed along the nephron?

$$H_2O \text{ reabsorbed} = 1 - \frac{1}{\left[\frac{TF}{P}\right]_{inulin}}$$

What equation gives the fraction of filtered load remaining at any point along the nephron?

$$\text{Fraction of filtered load} = \frac{\left[\frac{TF}{P}\right]_X}{\left[\frac{TF}{P}\right]_{inulin}}$$

where X represents any filtered substance

Give the percentage of Na⁺ reabsorption along the following parts of the nephron:

Part	%
Proximal convoluted tubule (PCT)	67%
Thin descending limb	0%
Thin ascending limb	0%
Thick ascending limb	25%
Distal convoluted tubule (DCT)	5%
Collecting ducts (CDs)	3%

What percentage of all of the filtered Na⁺ is excreted in a normal nephron — <1%

In the proximal tubule, what other substances are reabsorbed with Na⁺ and H₂O?
1. HCO_3^-
2. Glucose
3. Amino acids
4. Phosphate
5. Lactate

What percentage of glucose and amino acids is reabsorbed in the proximal tubule? — 100%

Describe the mechanism by which Na^+-H^+ exchange occurs.	Na^+ is reabsorbed by countertransport of H^+. This process is directly linked to HCO_3^- reabsorption
How is Na^+ reabsorbed in the middle and late tubules?	Na^+ is reabsorbed with Cl^-
Is Na^+ reabsorbed at a constant or variable rate in the proximal tubule?	Constant 67% of filtered Na^+
What is the name of the mechanism by which this is accomplished?	Glomerulotubular balance
What forces influence glomerulotubular balance?	Starling's forces in the peritubular capillary blood
What is the effect of the following on reabsorption:	
Increased GFR	Increased: by increase in peritubular [protein]
Increased FF	Increased: by increase in peritubular [protein]
ECF volume contraction	Increased: by increase in peritubular [protein] and decrease in peritubular pressure
ECF volume expansion	Decreased: by decrease in peritubular [protein] and increase in peritubular pressure
What transporter is responsible for reabsorption of Na^+ in the thick ascending limb of the loop of Henle?	Na^+-K^+-$2Cl^-$ cotransporter in the luminal membrane
What diuretics are responsible for inhibition of this transporter?	Loop diuretics: furosemide, ethacrynic acid, bumetanide
Is the thick ascending limb of the loop of Henle permeable to water?	No, therefore NaCl is reabsorbed without water
What happens to the osmolarity of the TF and [Na^+] in the thick ascending limb compared to plasma?	Decreases, which is why the segment is called the *diluting segment*
What is another name for the early distal tubule?	Cortical diluting segment
What transporter is responsible for Na^+ reabsorption in this segment?	Na^+-Cl^- cotransporter
What diuretics work on the transporter in this segment?	Thiazide diuretics
Is the early distal tubule permeable to water?	No

What happens to the osmolarity of the solution in the tubular lumen in this section?	It becomes further diluted
Name the cell types responsible for electrolyte transport in the late distal tubule and CDs.	1. Principal cells 2. α-Intercalated cells
What electrolytes are secreted and reabsorbed in principal cells?	Secreted: K^+ Reabsorbed: Na^+ and H_2O
Which cells does antidiuretic hormone (ADH) work on?	Principal cells
How does ADH work?	It directs the insertion of water channels (aquaporins) into principal cell luminal membranes
What effect does aldosterone have on electrolyte reabsorption or secretion?	1. Increases Na^+ reabsorption 2. Increases K^+ secretion
What percentage of overall Na^+ reabsorption is affected by aldosterone?	2%
What diuretics work on principal cells?	K^+-sparing diuretics (spironolactone, triamterene, amiloride), which decrease K^+ secretion
What electrolytes are secreted and reabsorbed in α-intercalated cells?	Secreted: H^+ (by H^+-ATPase) Reabsorbed: K^+ (by H^+-K^+-ATPase)
What influence does aldosterone have on α-intercalated cells?	Increases H^+ secretion by stimulating H^+-ATPase
Give the equation to calculate fractional excretion of sodium (Na^+) (FENa).	$FENa = 100\% \times \dfrac{Urine\ Na^+ \times Plasma\ Cr}{Plasma\ Na^+ \times Urine\ Cr}$ where urine concentrations are in mEq/L and plasma concentrations are in mg/dL
What conditions are associated with the following values of FENa:	
<1%	Acute glomerulnephritis, hepatorenal syndrome, states of prerenal azotemia (e.g., congestive heart failure or dehydration), and acute partial urinary tract obstruction
>1%	Acute tubular necrosis (ATN) and bilateral urethral obstruction

What are the values for FENa and urine [Na⁺] that are suggestive of prerenal azotemia or glomerulonephritis?	FENa <1% Urinary Na⁺ <20 mEq/L
What are the values for FENa and urine [Na⁺] that are suggestive of ATN or postrenal azotemia?	FENa >3% Urinary Na⁺ >40 mEq/L
Where is most of the body's K⁺ located?	ICF
Name some factors that lead to K⁺ entering cells.	1. Insulin 2. β-Adrenergic agonists 3. Alkalosis 4. Hyposmolarity
Name some factors that lead to K⁺ leaving cells.	1. Insulin deficiency (e.g., diabetes) 2. β-Adrenergic antagonists 3. Acidosis 4. Cell lysis 5. Exercise 6. Na⁺-K⁺ pump inhibitors 7. Hyperosmolarity
Give the percentage of K⁺ reabsorption along the various parts of the nephron:	
PCT	67%
Thin descending limb	0%
Thin ascending limb	20%
Thick ascending limb	Variable
DCT	Variable
CDs	Variable
What percentage of filtered K⁺ is excreted in a normal nephron?	1–110%
How is K⁺ reabsorbed in the thick ascending limb?	By the Na⁺-K⁺-2Cl⁻ cotransporter
How is K⁺ reabsorbed in the distal tubule and CDs?	By the H⁺-K⁺-ATPase in the luminal membranes of α-intercalated cells
What conditions cause K⁺ to be reabsorbed?	Low K⁺ diets (due to K⁺ depletion)
What is the mechanism of distal K⁺ secretion in the basolateral and luminal membranes?	Basolateral: Na⁺-K⁺ pump (active) Luminal: K⁺ channels (passive)
How do each of the following conditions affect K⁺:	
High K⁺ diet	Increase secretion
Low K⁺ diet	Decrease secretion

Hyperaldosteronism	Increase secretion
Hypoaldosteronism	Decrease secretion
Acidosis	Decrease secretion
Alkalosis	Increase secretion
Thiazide and loop diuretics	Increase secretion
Increased luminal anions	Increase secretion
Spironolactone, triamterene, amiloride	Decrease secretion (K^+ sparing diuretics)

How does aldosterone affect K^+ secretion in the kidney?

↑ aldosterone secretion
↓
↑ Na^+ entry across luminal membrane into the cells
↓
↑ activity of Na^+-K^+ pump on the basolateral membrane
↓
↑ Na^+ secretion out of the cells and ↑ K^+ cell entry across basolateral membrane
↓
↑ intracellular [K^+]
↓
↑ K^+ secretion across luminal membrane

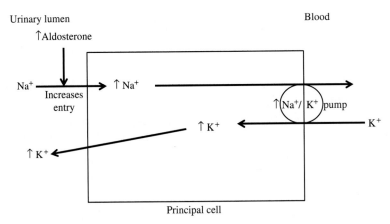

Figure 5.3 Aldosterone.

Renal and Acid-Base Physiology

Describe the pathway by which acidosis affects K^+ secretion.	↑ blood H^+ ↓ ↑ H^+ entry across basolateral membrane ↓ ↑ K^+ secretion across basolateral membrane ↓ ↓ intracellular K^+ ↓ ↓ K^+ secretion across luminal membrane
Describe the pathway by which alkalosis affects K^+ secretion.	↓ Blood H^+ ↓ ↓ H^+ entry across basolateral membrane ↓ ↓ K^+ secretion across basolateral membrane ↓ ↓ intracellular K^+ ↑ K^+ secretion across luminal membrane (reverse of acidosis)
Describe the mechanism by which thiazide and loop diuretics affect K^+ secretion.	↑ flow rate through distal tubule ↓ Dilution of luminal $[K^+]$ ↓ ↑ driving force for K^+ secretion
What diuretic is a direct antagonist to aldosterone?	Spironolactone
What diuretics work directly on the principal cells?	Triamterene and amiloride

OTHER ELECTROLYTES

Where is the majority of urea reabsorbed?	50% is passively reabsorbed in the proximal tubule
What parts of the kidney are impermeable to urea?	1. Distal tubule 2. Cortical CDs 3. Outer medullary CDs
On what part of the kidney does antidiuretic hormone work?	Inner medullary CDs

Where in the kidney is the majority of phosphate reabsorbed?	85% is reabsorbed in the proximal tubule (remaining 15% excreted in urine)
How is phosphate reabsorbed in the kidney?	By Na^+-phosphate cotransporter
How does parathyroid hormone affect phosphate regulation in the kidney?	Inhibits its reabsorption in the proximal tubule by inhibiting the Na^+-phosphate cotransporter via generation of cyclic AMP (cAMP)
What percentage of plasma Ca^{2+} is filtered across the glomerular capillaries?	60%
Where is the majority of Ca^{2+} reabsorbed?	Proximal tubule and thick ascending limb
What other electrolyte is coupled to Ca^{2+} reabsorption?	Na^+
What percentage of calcium reabsorption takes place in the distal tubule and CDs?	8%
What diuretics can be used in hypercalcemia?	Loop diuretics
What diuretics can be used in hypercalciuria?	Thiazide diuretics
What diuretics decrease Ca^{2+} excretion?	Thiazide diuretics
What parts of the kidney play a role in Mg^{2+} reabsorption?	1. Proximal tubule 2. Thick ascending limb 3. Distal tubule
What electrolyte competes with Mg^{2+} for reabsorption in the thick ascending limb?	Ca^{2+}
How does hypercalemia affect Mg^{2+} reabsorption?	Increases Mg^{2+} excretion

DILUTION AND CONCENTRATION OF URINE

In water deprivation, what part of the brain is activated?	Osmoreceptors in the anterior hypothalamus
What is the stimulus?	Increased plasma osmolarity
What part of the brain secretes ADH?	Posterior pituitary
What stimulates the secretion of ADH?	Anterior pituitary activation

Name some states associated with high levels of circulating ADH.	1. Hemorrhage 2. Water deprivation 3. Syndrome of inappropriate antidiuretic hormone (SIADH)
What is the corticopapillary osmotic gradient?	Osmolarity gradient from the cortex to the papilla
What are the primary electrolytes responsible for creating the corticopapillary osmotic gradient?	NaCl and urea
How is the gradient established?	Countercurrent multiplication and urea recycling
Describe the countercurrent multiplication in a kidney with high circulating ADH.	NaCl is reabsorbed in the thick ascending limb while there is countercurrent flow in the descending and ascending limbs of the loop of Henle
What hormone augments countercurrent multiplication in the kidney?	ADH: stimulates NaCl reabsorption in the thick ascending limb increasing the size of the gradient
What is urea recycling?	Process by which urea goes from the inner medullary CDs into the medullary interstitial fluid
What hormone augments urea recycling?	ADH
How is the urea gradient maintained in the kidneys?	Countercurrent exchange
What are the vasa recta?	Capillaries that supply the loop of Henle
How do vasa recta affect countercurrent exchange?	They serve as osmotic exchangers. The blood in them equilibrates osmotically with the interstitial fluid in the medulla and papilla
What parts of the nephron are impermeable to water?	Thick ascending limb of the loop of Henle and early distal tubule
What parts of the kidney increase reabsorption of water in the presence of ADH?	Late distal tubule and CDs
What happens to ADH with decreased plasma osmolarity?	Secretion is inhibited
Name some conditions that are associated with very low levels of circulating ADH.	Excess water intake and central diabetes insipidus

What is the osmolarity without ADH and with high levels of ADH in the following regions of the nephron:	High ADH	No ADH
Proximal tubule	300 mOsm/L	300 mOsm/L
Thick ascending limb	100 mOsm/L	120 mOsm/L
Early distal tubule	300 mOsm/L	300 mOsm/L
Late distal tubule	300 mOsm/L	100 mOsm/L
CDs	1200 mOsm/L	50 mOsm/L

What formula is used to estimate the ability to concentrate or dilute urine?

Free-water clearance (C_{H_2O})

$$C_{H_2O} = V - C_{osm}$$

where V = urine flow rate (mL/min) and C_{osm} = osmolar clearance ($U_{osm}(V/P_{osm})$)

What does a positive value for (C_{H_2O}) mean?

Excess water excretion: urine is hyposmotic to plasma (absence of ADH)

What does a negative value for (C_{H_2O}) mean?

Decreased water excretion: urine is hyperosmotic to plasma (high ADH)

ACID-BASE

What is the volatile acid used by the body?

CO_2

What enzyme is responsible for the following reaction:

$H_2O + CO_2 \rightarrow H^+ + HCO_3^-$?

Carbonic anhydrase

Name some nonvolatile acids used by the body.

1. Phosphoric acid
2. Ketoacids
3. Lactic acid

What are buffers?

Compounds that prevent changes in pH with the addition or removal of H^+

What generally makes the best buffers?

Weak acids and bases

When are buffers most effective in respect to their pK?

When they are within 1 pH unit of their pK

What is the major extracellular buffer? What is its pK?

HCO_3^-, pK = 6.1

What is a minor extracellular buffer? What is its pK?	Phosphate, pK = 6.8
What is the most important buffer in urine?	Phosphate
What does titratable acidity mean relative to the body?	Amount of NaOH required to neutralize a 24-hour urine specimen
What is the titratable acid used by the body?	$H_2PO_4^-$
Name some intracellular buffers.	Organic phosphates and proteins
Within the physiologic pH range, is deoxyhemoglobin or oxyhemoglobin a better buffer?	Deoxyhemoglobin
Write the Henderson-Hasselbalch equation for weak acids.	$pH = pK + \log \frac{[A^-]}{[HA]}$
Diagram the typical titration curve for a weak acid.	

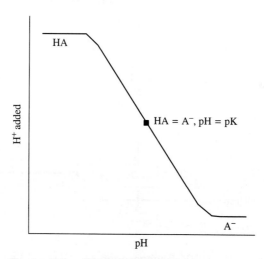

Figure 5.4 Weak Acid Titration.

Where on the curve is the buffer most effective?	On the linear portion of the curve (see the curve shown in Fig. 5.4)

What is important about the part of the graph where concentrations of HA and A⁻ are equal?	$pH = pK$
Where is the majority of filtered HCO_3^- reabsorbed?	Proximal tubule
Diagram the steps involved in HCO_3^- reabsorption.	

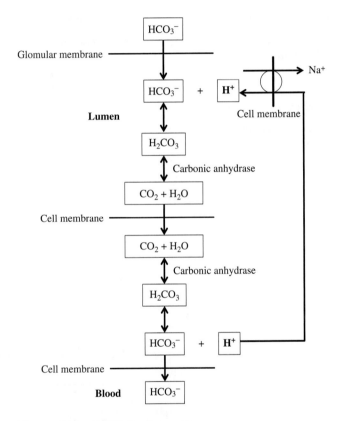

Figure 5.5 HCO_3^- Reabsorption.

In the process of HCO_3^- reabsorption, is there a net reabsorption or secretion of H^+?	Neither

What factors affect the reabsorption of filtered HCO_3^-?	1. Filtered load of HCO_3^- 2. P_{CO_2} 3. ECF volume 4. Angiotensin II
Describe how P_{CO_2} affects HCO_3^- reabsorption.	Increases in P_{CO_2}: increases rates of HCO_3^- reabsorption because the supply of intracellular H^+ increases Decreases in P_{CO_2}: decreases rates of HCO_3^- reabsorption because the supply of intracellular H^+ decreases
Describe how ECF volume affects HCO_3^- reabsorption.	↑ ECF volume: ↓ HCO_3^- reabsorption ↓ ECF volume: ↑ HCO_3^- reabsorption (also secretion of angiotensin II)
Describe how angiotensin II affects HCO_3^- reabsorption.	Stimulates the Na^+-H^+ pump, which leads to increase in H^+ secretion into the urine resulting in increased HCO_3^- reabsorption
How is H^+ secreted in urine?	As titratable acid and NH_4^+
Diagram the excretion of H^+ as titratable acid.	

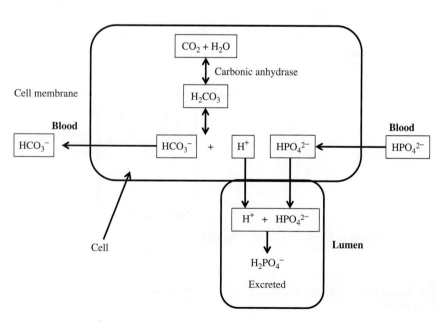

Figure 5.6 H^+ Excretion As Titratable Acid.

What factors influence H^+ excretion as titratable acid?	Amount of urinary buffer and pK of buffer
What factors influence H^+ excretion as NH_4^+?	NH_3 produced by the renal cells and urine pH
What do renal cells use to produce NH_3?	Glutamine
How does NH_3 get transported in the urinary lumen?	Down its concentration gradient by simple diffusion
What does the term *diffusion trapping* mean?	H^+ is secreted and combines with NH_3 to form NH_4^+, which *traps* the H^+ and allows for its excretion
What happens to excretion of NH_4^+ at lower pHs?	Increased excretion because there is a greater ratio of NH_4^+ compared to NH_3

Diagram the process of excretion of H^+ as NH_4^+.

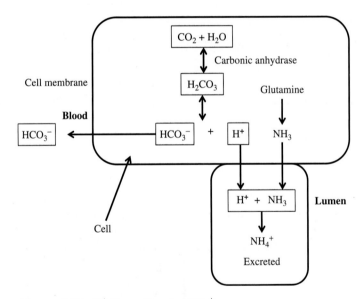

Figure 5.7 H^+ Excretion As NH_4^+.

What happens to NH_3 synthesis in acidosis?	Increases

Renal and Acid-Base Physiology

What is the primary disturbance in each of the following conditions:	
Metabolic acidosis	HCO_3^- deceases
Metabolic alkalosis	HCO_3^- increases
Respiratory acidosis	P_{CO_2} increases
Respiratory alkalosis	P_{CO_2} decreases
Describe the respiratory response in each of the following conditions:	
Metabolic acidosis	Hyperventilation
Metabolic alkalosis	Hypoventilation
Respiratory acidosis	None
Respiratory alkalosis	None
Describe the renal compensation in each of the following conditions:	
Metabolic acidosis	None
Metabolic alkalosis	None
Respiratory acidosis	↑ H^+ excretion and ↑ HCO_3^- reabsorption
Respiratory alkalosis	↓ H^+ excretion and ↓ HCO_3^- reabsorption
What is Kussmaul breathing?	Hyperventilation secondary to academia
What adaptation occurs in states of chronic metabolic acidosis?	Increase in NH_3 synthesis
What is the equation for serum anion gap (AG)?	$AG = [Na^+] - ([Cl^-] + [HCO_3^-])$
What does the value for AG represent?	Unmeasured anions in serum
Give some examples of unmeasured anions.	1. Phosphate 2. Citrate 3. Lactate 4. Sulfate 5. Formate 6. Methanol 7. Formaldehyde
What is the normal range for the AG?	8–12 mEq/L
In metabolic acidosis, what happens to chloride levels if the serum anion gap is normal?	Increases

What is the mnemonic for causes of anion gap metabolic acidosis?	**MUDPILES** Methanol Uremia (from chronic renal failure) Diabetic ketoacidosis (e.g., DKA) Paraldehyde/Phenformin Iron/Isoniazid (INH) Lactic acidosis Ethanol/Ethylene glycol Salicylates
List some conditions that result in nonanion gap metabolic acidosis.	1. Diarrhea 2. RTA types 1, 2, and 4
List some conditions that result in metabolic alkalosis.	1. Vomiting 2. Hyperaldosteronism 3. Loop and thiazide diuretics
If metabolic alkalosis occurs with ECF volume contraction, what happens to the reabsorption of HCO_3^-?	Increases and worsens the alkalosis!
How do diuretics cause metabolic alkalosis?	By volume contraction
How does vomiting lead to metabolic alkalosis?	H^+ is lost in vomitus and volume contraction
List some conditions that may result in respiratory acidosis.	1. Guillain-Barré syndrome 2. Polio 3. Amyotrophic lateral sclerosis (ALS) 4. Multiple sclerosis (MS) 5. Airway obstruction 6. Acute respiratory distress syndrome (ARDS) 7. Chronic obstructive pulmonary disease (COPD)
List some substances associated with respiratory acidosis.	1. Opiates 2. Sedatives 3. Anesthetics
List some conditions that result in respiratory alkalosis.	1. Pneumonia 2. Pulmonary embolus 3. High altitude 4. Psychogenic 5. Salicylate intoxication
In respiratory acidosis, what is the acute renal response?	There is none

In respiratory acidosis, what is the chronic renal response?	Increased HCO_3^- reabsorption and increased H^+ excretion
Describe how to calculate the compensation in each of the following conditions:	
Metabolic acidosis	1 mEq/L ↓ $[HCO_3^-]$ = 1.3 mm Hg ↓ P_{CO_2}
Metabolic alkalosis	1 mEq/L ↑ $[HCO_3^-]$ = 0.7 mm Hg ↑ P_{CO_2}
Acute respiratory acidosis	1 mEq/L ↑ $[HCO_3^-]$ = 0.1 mm Hg ↑ P_{CO_2}
Chronic respiratory acidosis	1 mEq/L ↑ $[HCO_3^-]$ = 0.4 mm Hg ↑ P_{CO_2}
Acute respiratory alkalosis	1 mEq/L ↓ $[HCO_3^-]$ = 0.2 mm Hg ↓ P_{CO_2}
Chronic respiratory alkalosis	1 mEq/L ↓ $[HCO_3^-]$ = 0.4 mm Hg ↓ P_{CO_2}

CHAPTER 6
Gastrointestinal Physiology

GASTROINTESTINAL TRACT

What is another name for the gastrointestinal (GI) tract?

Alimentary tract

Name the five main functions of the alimentary tract.

1. Motility
2. Secretion
3. Digestion
4. Absorption
5. Excretion

Please identify the labeled components of cellular anatomy in the following cross-section of the GI tract.

1. Lumen
2. Epithelial cells
3. Lamina propria
4. Muscularis mucosa
5. Submucosa
6. Circular muscle
7. Longitudinal muscle
8. Serosa

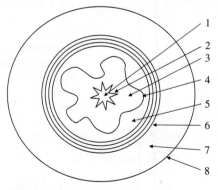

Figure 6.1 Cross-Section of Cellular Anatomy.

What is the main muscle type in the GI canal?	Visceral smooth muscle (VSM)
What is unique about the muscle in the GI tract?	The cells are interconnected by gap junctions and function together as a single unit. Thus, an action potential generated in one muscle cell can easily spread to adjacent cells, allowing the cells to contract as a single unit
What are some important factors about VSM?	1. Associated with the nervous plexus 2. Plasticity allows cells to respond to stretch with little tension 3. Electrical rhythm is transmitted throughout the luminal wall
Name the two main nervous systems of the gut.	1. Autonomic or extrinsic system 2. Enteric or intrinsic system
What are the components of the extrinsic system?	1. Parasympathetic nervous system (PNS) 2. Sympathetic nervous system (SNS)
Name the main function of the PNS.	Excitation
Which GI nerves carry PNS fibers and what structures/organs do they innervate?	Vagus: esophagus, stomach, pancreas, and upper portion of the large intestine Pelvic nerves: lower portion of large intestine, rectum, and anus
Is the action of PNS exclusively excitatory?	No
Where are the exceptions?	1. Inhibits somatostatin release in the antrum of the stomach 2. Inhibits fibers in the lower esophageal sphincter (LES), pyloric, and internal anal sphincter thus allowing them to open
Name the main function of the SNS.	Inhibitory
Which GI nerves carry SNS fibers?	Spinal cord
Is the action of SNS exclusively excitatory?	No
Where are the exceptions?	LES, pyloric, and internal anal sphincter

Gastrointestinal Physiology

Why does sympathectomy not have much of an effect on alimentary motility?	Reuptake of norepinephrine (NE) by sympathetic nerve endings is so rapid that only a great rise in NE concentration during a sympathetic discharge can have a significant effect on GI motility
What can overexcitation of the extrinsic nervous system produce?	Irritable bowel syndrome (IBS)
What are the components of the intrinsic system?	1. Myenteric or Auerbach's plexus 2. Submucosal or Meissner's plexus
What are the functions of the intrinsic system?	1. Acts as mediator of information between the extrinsic nervous system and the alimentary tract 2. Commands most functions of the GI especially motility and secretion 3. Can execute neural function of the gut without extrinsic innervations
Where is the myenteric plexus located?	It lies between the longitudinal and circular muscle layers
What is the main function of the myenteric plexus?	Controls motility
Where is the submucosal plexus located?	It lies in the submucosa (hence name), between the muscularis mucosa and circular muscle layer
Name its main function.	Controls secretions and blood flow
What can occur with the absence of the enteric nervous system in the colon?	Hirschsprung's disease (megacolon)
What are the three main types of GI reflexes?	1. Reflexes within the enteric nervous system 2. Reflexes that extend from the prevertebral sympathetic ganglion to the GI tract 3. Reflexes that extend from the spinal cord or brain stem to the GI tract
For the following description, name the specific reflex:	
Initiated by distention of stomach and destroyed by a vagotomy	Vagovagal reflex
Associated with gastric and intestinal phases of pancreatic exocrine secretion	Vagovagal reflex

Increases peristalsis in the terminal ileum due to the food in the stomach	Gastroileal reflex
Inhibits stomach's motility and secretion due to distention of ileum	Ileogastric reflex
Increases motility and evacuation of proximal and distal colon due to food in the stomach	Gastrocolic reflex
Distention in one segment of colon causes relaxation in another segment of colon	Colonocolic reflex
Mediated by gastrin	Gastroileal and gastrocolic reflex

GASTROINTESTINAL HORMONES

Name the main stimuli for hormone release.	Food in the gut
Name the main function of GI hormones.	Regulate the digestive process by influencing secretion, motility, and blood flow
Name the four systems used by hormones to exert their functions.	1. Neurocrine 2. Paracrine 3. Endocrine 4. Neuroendocrine
Define neurocrine.	Process in which one nerve fiber releases messenger that acts across a short distance upon a target cell (nerve fiber, muscle cell, or gland cell)
Define paracrine.	Released messenger acts upon adjacent cells
Define endocrine.	Stimulus acting upon a receptor causes the cell to release a messenger into the blood that acts on a distant target cell
Define neuroendocrine.	Action potential causes release of messenger that acts upon a distant target cell
Name the main GI hormones.	1. Gastrin 2. Cholecystokinin (CCK) 3. Secretin 4. Gastric inhibitory peptide (GIP)

Gastrointestinal Physiology

What are the three forms of gastrin?	1. Big gastrin (34 amino acids) 2. Little gastrin (17 amino acids) 3. Mini gastrin (14 amino acids)
Which form is most abundant and potent?	Little gastrin
Which form has the longest $t_{1/2}$?	Big gastrin (42 minutes)
On which amino acids is the physiological activity located?	Last four amino acids at the carboxy terminal (i.e., little gastrin: AA 14-17)
Name the cells that secrete gastrin.	G cells in the antral mucosa of stomach
What are the functions of gastrin?	1. Increases hydrochloric acid (HCl) secretion (via parietal cells) 2. Stimulates growth of gastric mucosa 3. Increases gastric motility 4. Increases LES pressure (preventing reflux) 5. Lowers ileocecal sphincter pressure (allows defection) 6. Increases pepsinogen secretion
What are the major stimuli for gastrin's secretion?	1. L-Amino acids (i.e., phenylalanine, tryphtophan, cystein, tyrosine) 2. Vagal stimulation 3. Stomach distention
What are some other stimuli for gastrin's secretion?	1. Epinephrine 2. Calcium 3. Acetylcholine (ACh)
Name the inhibitors of gastrin secretion.	1. pH <2 2. Somatostatin 3. Secretin 4. Calcitonin 5. GIP 6. Glucagon 7. Vasoactive inhibitory peptide (VIP)
What system is primarily used by gastrin to exert its actions?	Endocrine
Which other GI hormone is "related" to gastrin?	CCK, which shares five amino acids on the carboxy terminal
What functions make them related?	Both stimulate gastric acid and pancreatic enzyme secretion

How do they differ in this shared function?	Potency
What syndrome occurs when non-β-cell tumors of the pancreas secrete gastrin (e.g., gastrinoma)?	Zollinger-Ellison syndrome
What are the three forms of CCK?	1. CCK 39 2. CCK 33 3. CCK 8
Which form is most abundant and potent?	CCK 8 (octapeptide)
On which amino acid sequence is the physiological activity located?	On the octapeptide on the carboxy terminal
Name the cells that secrete CCK.	I cells of duodenum and jejunum
What are the functions of CCK?	1. Increase gallbladder contraction 2. Lower pressure on the sphincter of Oddi 3. Increase pancreatic enzyme secretion 4. Decrease gastric emptying rate 5. Increase pepsinogen secretion 6. Lower LES pressure 7. Stimulate growth of the exocrine pancreas 8. Work synergistically with secretin to increase bicarbonate secretion
What are the major stimuli for CCK's secretion?	1. Small peptides and L-amino acids 2. Fatty acids and monoglycerides
Why don't triglycerides stimulate the release of CCK?	They cannot cross the intestinal membranes
Which systems are primary used by CCK to exert its actions?	Endocrine and neurocrine
What size polypeptide is secretin?	27 amino acid polypeptide
What is secretin's claim to fame?	It was the first GI hormone to be discovered (Bayliss and Starling, 1902)
On which amino acid sequence is the physiological activity located?	All 27 amino acids are needed
Name the cells that secrete secretin.	S cells of the duodenum

Gastrointestinal Physiology

What are the functions of secretin?	1. Stimulates bicarbonate secretion from pancreatic and biliary duct cells 2. Enhances activity of CCK on pancreatic secretion and gallbladder contraction 3. Lowers gastric and intestinal motility 4. Lowers HCl secretion 5. Increases pepsinogen
What are the major stimuli for secretin's release?	1. pH <4.5 in duodenum 2. Fatty acids, amino acids, and peptides in the duodenum
What system is primarily used by secretin to exert its actions?	Endocrine
Which other GI hormone is "related" to secretin?	Glucagon (they share 14 amino acids)
Which hormones are part of the secretin-glucagon family?	1. Secretin 2. Glucagon 3. Vasoactive inhibitory peptide (VIP) 4. GIP
What size polypeptide is GIP?	42 amino acid polypeptide
Name the cells that secrete GIP.	K cells of duodenum and jejunum
What are the functions of GIP?	1. Stimulates insulin release 2. Inhibits H^+ secretion
What are the major stimuli for GIP's release?	1. Fat 2. Protein 3. Carbohydrate
What system is primarily used by GIP to exert its actions?	Endocrine
Which other GI hormone(s) are related to GIP?	Secretin and glucagon
What are the GI paracrines?	Somatostatin and histamine
Name the cells that secrete somatostatin.	Multiple cells in the GI tract
What is the stimulus for somatostatin release?	Presence of H^+ in the lumen

What inhibits the secretion of somatostatin?	Vagal stimulation
What is the function of somatostatin?	1. Inhibits release of all GI hormones 2. Inhibits gastric H^+ secretion
Name the cells that secrete histamine.	Mast cells within the gastric mucosa
What is the function of histamine in the GI tract?	Increases gastric H^+ secretion (both directly and by potentiation of the effects of gastrin and vagal stimulation)
What are the GI neurocrines?	1. VIP 2. Gastrin-releasing peptide (GRP) (bombesin) 3. Enkephalins
What size polypeptide is VIP?	28 amino acid polypeptide
What other GI hormone is VIP homologous to?	Secretin
Name the cells that normally secrete VIP.	Neurons in the mucosa and smooth muscle of the GI tract
What tumor type can also secrete VIP?	Pancreatic islet cell tumors
What are the functions of VIP?	1. Relaxes GI smooth muscle (including LES) 2. Stimulates pancreas to secrete HCO_3^- 3. Inhibits gastric H^+ secretion
Name the cells that secrete GRP.	Vagal nerves that innervate G cells
What is the function of GRP?	Stimulates gastrin release
What are the types of enkephalins?	1. Met-enkephalin 2. Leu-enkephalin
Name the cells that secrete enkephalins.	Neurons in the mucosa and smooth muscle of the GI tract
What are the functions of enkephalins?	1. Contract GI smooth muscle (especially lower esophageal, pyloric, and ileocecal sphincters) 2. Inhibit secretion of fluid and electrolytes by the intestines

GASTROINTESTINAL MOTILITY

Name the types of electrical waves found in the alimentary tract.	1. Slow waves 2. Spike potentials

Gastrointestinal Physiology

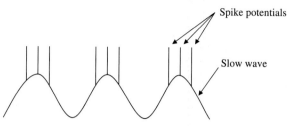

Figure 6.2 Electrical Waves.

What are slow waves?	Fluctuating changes in the resting membrane potential
What are they not?	Action potentials
Where are they generated?	Cells of Cajal (pacemaker of the alimentary tract)
Why are they important?	Determine the rhythmicity of the GI tract's contractions by controlling the pattern of spike potentials
Where in the tract are the waves the fastest?	Duodenum at 12 waves/min
Where in the tract are the waves the slowest?	Stomach at 3 waves/min
What are spike potentials?	Action potentials of the alimentary tract
How are they generated?	They occur when the resting membrane potentials becomes greater than −40 mV (e.g., membrane depolarizes)
Name three factors that cause depolarization of the membrane.	1. Muscle stretch 2. ACh 3. PNS
Which channels are involved in the generation of the action potential?	Ca^{2+}-Na^+ channels
How does a spike potential cause contraction?	Ca^{2+} enters the cell
Define motility.	Mechanical activity of the GI tract that is divided into mixing (segmentation) and propelling (peristalsis)
Describe segmentation.	Contraction around the bolus sends intestinal contents (chyme) backward and forward. The area then relaxes and the material moves back into the segment, mixing the contents

Describe peristalsis.	Contraction behind the bolus is coupled with relaxation in front of it, which propels the bolus down the GI tract
Where is the swallow center located?	Medulla and lower pons
Which nerves contain the motor impulses from the swallow center?	Cranial nerves (CN) V, IX, X, XII, and the superior CN
Name the stage of the swallow reflex that is described by the following:	
Involves voluntary action that squeezes food into the pharynx and against the palate	Voluntary stage (oral)
Involves closure of the trachea, opening of the esophagus, and generation of a peristaltic wave (primary peristalsis) that forces the food bolus into the esophagus	Pharyngeal stage
Involves continuation of primary peristalsis, the relaxation of the upper esophagus, and the entrance of the food bolus into the esophagus	Esophageal stage
How long does it take the primary wave to reach the LES?	5–10 seconds (travels at 3–5 cm/s)
When is a secondary peristalsis generated?	When the primary peristalsis wave is insufficient to clear the esophagus of the food bolus
What is receptive relaxation?	A vagovagal reflex that relaxes the LES prior to the peristaltic wave
Why is receptive relaxation important?	Allows for easy propulsion of food bolus into the stomach
What is the mechanism by which gastric contents are able to reflux into the esophagus (e.g., gastroesophageal reflux disease or GERD)?	Decreased LES tone
What can result if the LES tone is increased and it does not relax with swallowing?	Achalasia
Name the anatomic divisions of the stomach in the following figure:	1. Cardia 2. Fundus 3. Body 4. Antrum 5. Pylorus

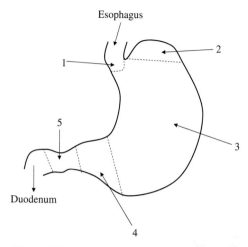

Figure 6.3 Anatomy of Stomach.

What are the physiologic divisions of the stomach and what are their boundaries?	Orad portion: extends from the fundus to the proximal body Caudad portion: extends from the antrum to the distal body
What are the three motor functions of the stomach?	1. Store food 2. Make chyme 3. Empty food at a rate suitable for proper digestion and absorption by the small intestine
What is the maximum amount of food that can be stored by the stomach?	1.5 L
What is chyme?	Semifluid paste that results from the food bolus mixing with gastric secretions
What promotes gastric emptying?	Stretch and gastrin
What inhibits gastric emptying?	1. Increased osmolarity 2. Products of fat digestion 3. pH <3.5
Which factors are involved in the inhibitory process?	Ileogastric reflex and CCK
Name the functions of small intestinal motility.	1. Allows for mixing of food bolus with digestive enzymes 2. Exposes food molecules to absorptive mucosa 3. Propels nonabsorbed material to the colon

Which aspect of motility is most important in the small intestine?	Segmentation: allows for increased surface area for digestion and absorption of chyme
What is the frequency of slow waves in the following segments of the small intestine:	
Duodenum	12 waves/min
Proximal jejunum	12 waves/min
Terminal ileum	8–9 waves/min
What other factor is important for segmental contraction?	Excitation by the myenteric plexus
What is the average velocity of peristalsis waves in the small intestine?	0.5–2.0 cm/s
What factors stimulate increased peristalsis activity?	1. Gastroileal reflex 2. Gastrin 3. CCK 4. Serotonin 5. Insulin
What factors inhibit peristalsis activity?	Secretin and glucagon
What is the principal function of the ileocecal valve?	Prevents backflow of colonic contents into the small intestine
Name the two types of motility found in the colon.	1. Haustrations 2. Mass movement
What are haustrations?	Combined contractions of the circular and longitudinal muscles of the large intestine that occur from the ileocecal valve to the transverse colon
How long does it take to move chyme from the ileocecal valve to the transverse colon?	8–15 hours
What are mass movements?	Modified peristalsis that is characterized by uniform contraction and movement of colonic contents down the descending colon
How often does it occur per day?	1–3 times/day
Name some factors that stimulate mass movement.	1. Gastrocolic reflex 2. Duodenocolic reflex 3. Irritation of the colon 4. PNS stimulation 5. Over distention of a colonic segment

Gastrointestinal Physiology

What initiates the desire for defecation?	Presence of feces in the rectum
Name the defecation reflex described by the following:	
Uses enteric nervous system to initiate peristaltic waves in the descending colon, sigmoid, and rectum, thus forcing feces toward the anus	Intrinsic reflex
Uses parasympathetic fibers in the pelvic nerves to intensify the peristaltic wave, relax the internal anal sphincter, and converts the intrinsic defecation reflex into a powerful process.	Parasympathetic defecation reflex
Where is the vomiting center located?	Medulla
What stimuli does the vomiting center respond to?	1. Gag 2. Gastric distention 3. Vestibular stimulation
Where are the chemoreceptors that can induce vomiting?	Fourth ventricle
What stimuli do the chemoreceptors respond to?	1. Emetic substances 2. Radiation 3. Vestibular stimulation
What is vomiting?	Reverse peristalsis that propels GI contents in the orad direction and out through the upper esophageal sphincter
What occurs if the peristalsis is not strong enough to overcome the pressure in the upper esophageal sphincter (UES)?	Retching
Where does the reverse peristalsis begin?	Small intestine

GASTROINTESTINAL SECRETIONS

What are the principal glands of salivation?	1. Parotid 2. Submandibular 3. Sublingual
Name the functions of saliva.	1. Moistens the mouth 2. Aids speech 3. Dissolves and buffers ingested food particles 4. Protects the oral cavity

Name and describe the two types of salivary secretions.	Serous secretions: contain enzymes for starch digestion Mucous secretions: contain mucin for lubrication and protection
Name the type of secretions for the principal glands:	
Parotid	Serous
Submandibular	Serous and mucous
Sublingual	Serous and mucous
What is an acinus?	Blind end of each duct in the salivary glands
What cells eject saliva from the glands into the mouth?	Myoepithelial cells
Describe the process of saliva production.	Initial saliva (isotonic to plasma) ↓ Ducts secrete K^+ and HCO_3^- ↓ Aldosterone ↑ Na^+ reabsorption and K^+ secretion (hypotonic saliva)
Why is a high flow rate of saliva associated with saliva that resembles the initial secretion?	There is decreased contact time for the secretion to be modified
What regulates saliva production?	PNS and SNS
What effect does the PNS have on saliva production?	Increases it
What effect does the SNS have on saliva production?	Increases it as well
What stimuli increase saliva production?	1. Presence of food in the mouth 2. Smells 3. Conditioned reflexes (e.g., Pavlov's dog) 4. Nausea
What stimuli decrease saliva production?	1. Dehydration 2. Fear 3. Anticholinergic medications 4. Sleep
What types of secretions are found in the esophagus?	Mucoid
Name its main function.	Provides lubrication for swallowing
What are the two types of gastric secretion glands?	1. Gastric (oxyntic) gland 2. Pyloric gland

Gastrointestinal Physiology

Name the cell types with the following description:

Found in the fundus and secretes HCl and intrinsic factor	Parietal cells
Found in the fundus and secretes pepsinogen	Chief cells/peptic cells
Found in antrum and secrets gastrin	G cells
Found in the antrum and secretes mucus and pepsinogen	Mucous cells

Schematically describe the mechanism of HCl secretion.

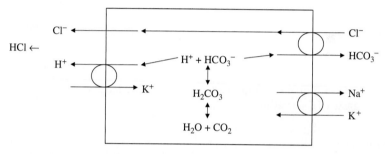

Figure 6.4 H^+ Secretion Mechanism.

Name the three factors that work synergistically to promote HCl secretion.

1. Histamine
2. Gastrin
3. Acetycholine

Schematically describe their relationship

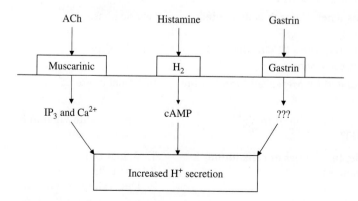

Figure 6.5 H^+ Secretion Stimulation.

What induces ACh stimulation of H^+ secretion?	Vagus nerve, which directly innervates parietal cells (ACh is the neurotransmitter)
What induces gastrin stimulation of H^+ secretion?	1. Small peptides present in the lumen 2. Distension of the stomach 3. Vagal stimulation (e.g., response to eating)
Name the compounds that can inhibit the following sources of HCl secretion:	
ACh	Cholinergic muscarinic antagonists
Histamine	H_2 receptor-blocking agents
Gastrin	None known at this time
H^+-K^+-ATPase	Proton pump inhibitors (PPIs)
What inhibits gastric H^+ secretion?	pH <3.0 in the stomach and the presence of chyme in the duodenum
What type of mechanism is employed to inhibit gastric H^+ secretion?	Negative feedback
Name the phase of gastric secretion that corresponds to the following description:	
Occurs before food enters the stomach and results from the sight, smell, and thought of food	Cephalic phase
Mediated entirely by vagal impulses	Cephalic phase
Accounts for 20–30% of the total gastric secretion	Cephalic phase
Mediated by vagovagal reflex, enteric reflexes, and gastrin	Gastric phase
Begins when stomach is distended by food	Gastric phase
Accounting for > 70% of the total gastric secretion	Gastric phase
Stimulated by presence of food in the duodenum	Intestinal phase
What is the function of intrinsic factor (IF)?	Complexes with vitamin B_{12} to allow for its absorption
Name the three types of cells found in the pancreas.	1. Acinar cells 2. Islet cells 3. Duct cells

Gastrointestinal Physiology

For the following secretion, name its cell of origin:

Zymogens (proteolytic enzymes)	Acinar cells
Glucagon	Islet cells (alpha cells)
Insulin	Islet cells (beta cells)
Somatostain	Islet cells (delta cells)
Isosmotic sodium bicarbonate solution	Duct cells

Name the enzyme with the following catalytic function:

Cleaves peptide bonds adjacent to basic amino acids	Trypsin
Cleaves peptide bonds adjacent to aromatic amino acids	Chymotrypsin
Cleaves carboxy terminal amino acids with aromatic or branched side chains	Carboxy peptidase A
Cleaves carboxy terminal amino acids with basic side chains	Carboxy peptidase B
Cleaves bonds in elastin	Elastase
Cleaves phospholipid to fatty acids and lysophospholipids	Phospholipase A
Hydrolyzes starches, glycogen, and other carbohydrates	Amylase

What substances stimulate the acinar cells?	Acetylcholine and CCK
Which substances stimulate the ductal cells?	Secretin
Which alimentary tract organ produces bile?	Liver
Where is bile stored?	Gallbladder
Name the two main functions of bile.	1. Helps with fat digestion and absorption 2. Helps excrete waste products such as bilirubin
Name the main components of bile.	1. Bile salts 2. Lecithin 3. Bilirubin 4. Cholesterol
What are bile salts?	Products of cholesterol and the conjugated amino acids: glycine and taurine (i.e., taurocholic)

Name the two functions of bile salts.	1. Act as emulsifier with lecithin 2. Help with the absorption process for fats
What is the most potent stimulus for gallbladder contraction?	CCK
Name some other less potent stimuli.	Acetylcholine and fatty food in the duodenum
Name the three stimuli that relax the sphincter of Oddi.	1. CCK 2. Contraction of gallbladder 3. Peristaltic waves of the intestine (most powerful)
What is enterohepatic circulation?	Recycling process for bile salts via active transport from distal ileum → portal circulation → liver
What is the function of Brunner's gland?	Secretes mucous into the small intestine
Where are they located?	In the first few centimeters of the duodenum between the pylorus and the ampula of vater
What is the function of the crypts of Lieberkuhn?	Secretes mucous into both the small and large intestine
What are the stimuli for mucous secretion?	Tactile stimuli and vagal stimulation
Name the main function of mucous.	Protects the walls of the intestine

GASTROINTESTINAL DIGESTION

Name three major sources of carbohydrate in the diet.	1. Starch 2. Sucrose 3. Lactose
What is the final product of carbohydrate digestion?	Monosaccharides
Name the major monosaccharide.	1. Glucose (80%) 2. Galactose 3. Fructose
Name the location where the following enzymes can be found:	
α-Dextrinase	Intestinal brush border
Isomaltase/sucrase	Intestinal brush border
Lactase	Intestinal brush border
Maltase	Intestinal brush border

Ptylain (alpha-amylase)	Saliva
Pancreatic amylase	Pancreatic secretions
Trehalase	Intestinal brush border

Name the enzyme used to digest the following:

Starch	Ptylain and pancreatic amylase
Lactose	Lactase
Sucrose	Sucrase
Maltose	Maltase
Alpha-limit dextrin	α-Dextrinase

Name the end product of digestion for the following:

Starch	Maltose and small glucose polymers (i.e., maltotriose, isomaltose, and alpha-limit dextrin)
Lactose	Glucose and galactose
Sucrose	Glucose and fructose
Maltose	Glucose
Alpha-limit dextrin	Glucose

Name the sources of proteins targeted for digestion.

1. Tissue protein
2. Dietary protein
3. Sloughed cells of GI tract
4. Pancreatic secretions

Name the location for the following enzymes of protein digestion:

Carboxy polypeptidase	Pancreatic secretions
Chymotrypsin	Pancreatic secretions
Pepsin	Stomach
Peptidases	Intestinal brush border
Proelastase	Intestinal brush border
Trypsin	Pancreatic secretions

What are the final products of protein digestion?

1. Amino acids (99%)
2. Dipeptides/tripeptides (<1%)

What *fats* are found in the diet?

1. Triglycerides
2. Cholesterol esters
3. Phospholipids
4. Cholesterol
5. Fat-soluble vitamins (A, D, E, and K)

Which is the most abundant?	Triglycerides
Where is the main site of fat digestion?	Small intestine
What makes fat digestion different from the digestion of carbohydrates and proteins?	The need for emulsification
Why is emulsification important?	Increases surface area for which digestive enzymes can operate
What is required for emulsification?	Bile salts and lecithin to form micelles
Why are micelles important in digestion?	Act as transporter for digested fat to the intestinal brush border, which helps the digestive process proceed without interruption
Name the enzyme used to digest the following:	
Triglycerides	Pancreatic lipase
Cholesterol ester	Cholesterol esterase
Phospholipid	Phospholipase A2
Name the end product of digestion for the following:	
Triglycerides	Free fatty acids and 2-monoglycerides
Cholesterol ester	Free fatty acid and cholesterol
Phospholipid	Free fatty acid and lysophospholipid

GASTROINTESTINAL ABSORPTION

The basic mechanism of absorption are	1. Solvent drag (movement of solute with water) 2. Passive diffusion 3. Facilitated diffusion 4. Active transport
What are the three anatomic additions to the small intestine that increases its absorption capacity 1000-fold?	1. Valvulae conniventes (folds of Kerckring) 2. Villi 3. Microvilli
What is the main mechanism used to absorb water?	Diffusion

Gastrointestinal Physiology

How much water is absorbed per day?	Between 5 L and 10 L
What is the average rate of water absorption by the small intestine?	200–400 mL/h
Name the mechanisms used to absorb sodium.	1. Simple diffusion 2. Facilitated diffusion 3. Solvent drag 4. Active transport 5. Coupled transport
Which is the most important mechanism for sodium absorption?	Coupled transport
Give examples of nutrients coupled to sodium's absorption.	Chloride, amino acids, and glucose
Name the potent stimuli for sodium's absorption.	Aldosterone
Name the mechanism(s) used for chloride's absorption.	Passive diffusion and exchange for HCO_3^-
What is the importance of the exchange with bicarbonate?	Bicarbonate is used to neutralize acid products
Describe the process of calcium absorption.	Calcium-binding protein binds calcium in the lumen, which enters the cell via facilitated diffusion. It is then actively transported into the interstitial fluid
What are the regulatory hormones for this process?	1. Parathyroid hormone (PTH) 2. 1, 25-Dihydroxycholecalciferol (1, 25-DHC)
Name the hormone whose action corresponds to the following:	
Stimulates the synthesis of calcium-binding protein	1, 25-DHC
Stimulates the formation of physiologically active vitamin D	PTH
Increases the active transport of calcium to the interstitial fluid	PTH and 1, 25-DHC
The apical transporter for iron is coupled to which enzyme?	Ferric reductase
The basolateral transporter for iron is coupled to which enzyme?	Ferroxidase

Why are ferric reductase and ferroxidase important?	Help change the oxidation state of iron, allowing for absorption
Name the main mechanism for glucose and galactose's absorption.	1. Sodium glucose/galactose cotransporter. 2. Sodium-dependent glucose transporter (SGLT-1)
Describe the process of glucose and galactose absorption.	Active transport of sodium through the basolateral membrane creates a gradient and a desire for sodium to diffuse through the apical membrane. The gradient allows sodium to combine with a transport protein that transports glucose/galactose into the cell
Name the other glucose transport mechanism.	Solvent drag (at high concentrations of glucose)
How is fructose absorbed?	Facilitated diffusion utilizing sodium-independent insulin-independent transporter (GLUT-5)
Where are most protein peptides absorbed?	Through the luminal membranes of the intestinal epithelial cells
How are most protein peptides absorbed?	Sodium-dependent cotransporter system
Where in the GI tract are most fats absorbed?	Jejunum
How are they transported to the brush border for absorption?	Bile acid micelles
What percentage of fat is absorbed with bile acid micelle?	97%
What percentage of fat is absorbed without bile acid micelle?	40–50%
What are chylomicrons?	Globules of reformed/repackaged triglycerides, cholesterol, and phospholipids that are arranged in a manner similar to micelles
What is the role of chylomicron formation?	Allows for systemic absorption of fat via exocytosis
What is essential for the exocytosis of chylomicrons.	Apoprotein B: helps the chylomicron attach to cell membrane before it is expelled

What are the fat-soluble vitamins?	Vitamins A, D, E, and K
What structure facilitates the absorption of fat-soluble vitamins?	Micelles
How are water-soluble vitamins absorbed?	Na^+-dependent cotransport
What is unique about vitamin B_{12} absorption?	It requires intrinsic factor (IF)
Where is vitamin B_{12} absorbed?	Terminal ileum
What results when there is a lack of IF?	Pernicious anemia (megaloblastic)

CHAPTER 7
Endocrine and Reproductive Physiology

HORMONES

What is a hormone?	A chemical substance, formed in a tissue or organ and carried in the blood, that stimulates or inhibits the growth or function of one or more other tissues or organs
What is an endocrine pathway?	A hormone secreted into blood that acts on distant target cells
What is a paracrine pathway?	A hormone released from one cell that acts on neighboring cells
What is an autocrine pathway?	A hormone released that acts on the cell that secreted it
How are polypeptide hormones synthesized?	Preprohormone produced (rough endoplasmic reticulum [ER]) ↓ cleaved Prohormone ↓ cleaved Hormone (Golgi apparatus) ↓ Packaged into granules for release
How are steroid hormones synthesized?	**Cholesterol** ↓ Pregnenolone (mitochondria) ↓ Side chain modifications (ER) ↓ Various hormones

How are amino acid hormones synthesized?	Tyrosine ↓ hydroxylation ↓ decarboxylation Dopamine ↓ further modification Various hormones
What is the fundamental mechanism of all hormone action?	Reversible, noncovalent binding to specific receptors on or in target cell
Where are polypeptide hormone receptors?	On the target cell
Where are steroid hormone receptors?	In the target cell cytoplasm
Where are amino acid hormone receptors?	On (catecholamines) or in (thyroid hormones) the target cell
What are G-proteins?	Guanosine 5'-triphosphate (GTP)-binding proteins that couple hormone receptors on the cell surface to a second messenger system inside the cell
What type of intrinsic property do G-proteins have?	GTP-ase activity
What type of G-proteins are there?	Either stimulatory (G_s) or inhibitory (G_i)
Describe the mechanism of the cyclic adenosine monophosphate (cAMP) second messenger system.	Hormone binds G-protein coupled receptor ↓ Activates adenylate cyclase ↓ ↑ cAMP ↓ ↑ protein kinase A phosphorylation of proteins ↓ Activation/inhibition of a metabolic process

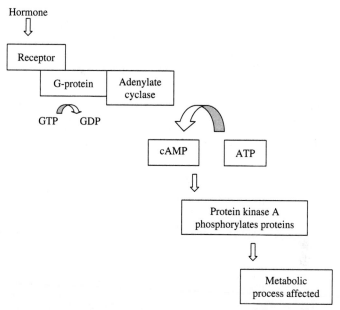

Figure 7.1 cAMP Pathway.

Describe the mechanism of the inositol triphosphate (IP₃) second messenger system.

Hormone binds G-protein coupled receptor
↓
Activates phospholipase C
↓
Frees diacylglycerol (DAG) + IP$_3$ from membrane
↓
Ca^{2+} release from ER
↓
Activates protein kinase C phosphorylation of proteins
↓
Activation/inhibition of a metabolic process

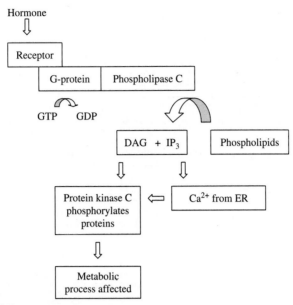

Figure 7.2 IP$_3$ Pathway.

Describe the mechanism of the intracellular Ca^{2+} second messenger system.	Hormone binds G-protein coupled receptor ↓ Activates membrane Ca^{2+} channel or releases Ca^{2+} from ER ↓ ↑ intracellular Ca^{2+} ↓ ↑ Ca^{2+}-calmodulin complex ↓ Regulation of other enzyme activities

Endocrine and Reproductive Physiology

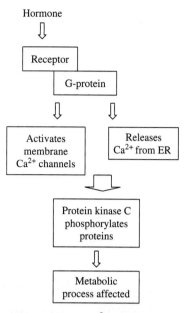

Figure 7.3. Ca^{2+} Pathway.

Describe the mechanism of the steroid and thyroid hormones.

Hormone crosses cell membrane
↓
Binds to receptor in cytoplasm
↓
Hormone-receptor complex enters nucleus
↓
Exposes DNA-binding domain on receptor
↓
Domain interacts with DNA to initiate transcription
↓
Protein synthesized with subsequent physiologic actions

What are the two principles of hormone receptor regulation?

1. Down-regulation: ↓ number or affinity of receptor for a hormone
2. Up-regulation: ↓ number or affinity of receptor for a hormone

What are the two principles of hormone secretion regulation?

1. Negative feedback (most common)
2. Positive feedback (rare)

What is negative feedback?	A hormone's actions directly or indirectly inhibit its own secretion—a self-terminating cycle
What is positive feedback?	A hormone's actions directly or indirectly promote its own secretion—a self-perpetuating cycle

HYPOTHALAMUS AND PITUITARY GLAND

Name the hormones of the:

Anterior pituitary	1. Thyroid stimulating hormone (TSH) 2. Luteinizing hormone (LH) 3. Follicle stimulating hormone (FSH) 4. Growth hormone (GH) 5. Prolactin (PRL) 6. Adrenocorticotropic hormone (ACTH)
Posterior pituitary	1. Oxytocin 2. Antidiuretic hormone (ADH)

Name the hypothalamic hormones and their function:

Thyrotropin-releasing hormone (TRH)	↑ TSH, PRL secretion
Gonadotropin-releasing hormone (GnRH)	↑ LH, FSH secretion
Corticotropin-releasing hormone (CRH)	↑ ACTH secretion (and α-[MSH], β-endorphin)
Growth hormone-releasing hormone (GHRH)	↑ GH secretion
Somatostatin (SS)	↓ release of GH, TSH (among others)
Prolactin inhibitory factor (PIF) (aka dopamine)	↓ release of PRL

Name the connection between the hypothalamus and the following:

Anterior pituitary	Hypothalamic-hypophysial portal system
Posterior pituitary	Hypothalamic tract

What is the hypothalamic-hypophysial portal system?	Capillaries that carry blood from the hypothalamus to the anterior pituitary and from the anterior pituitary back to the hypothalamus
What is the significance of the retrograde blood flow?	Feedback
How are the anterior pituitary hormones categorized?	1. GH-related hormones 2. Glycoprotein hormones 3. Corticotropin-related hormones

What hormones are included in the following hormone categories:

GHRH	1. GH 2. PRL 3. Human placental lactogen (HPL)—from placenta 4. Insulin-like growth factor (IGF)—from liver
Glycoprotein hormones	1. LH 2. FSH 3. TSH 4. Human chorionic gonadotropin (hCG)—from placenta
CRH	1. ACTH 2. MSH 3. Endorphins 4. Enkephalins 5. Lipotropins

What is unique about the makeup of the following:

GH-related hormones	GH is a polypeptide and is homologous with PRL and HPL
Glycoprotein hormones	All contain α- and β-subunits—α-subunits are similar; hormonal activity comes from β-subunits
CRH	All are from the same precursor proopiomelanocortin (POMC)

Diagram POMC proteolytic processing.

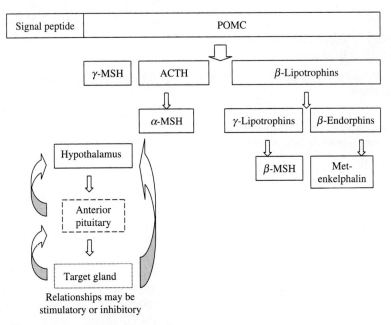

Figure 7.4 POMC Processing.

Name the actions of the anterior pituitary hormones:

TSH	↑ T_3 and T_4 production (see thyroid section)
LH	↑ estrogen, androgen production
FSH	↑ oocyte and sperm maturation (see ovary and testis sections)
GH	↑ general growth ↓ glucose uptake into cells → diabetogenic ↑ protein synthesis, ↑ lipolysis ↑ IGF production in liver
PRL	↑ milk production ↑ breast development Inhibits ovulation via ↓ GnRH
ACTH	↑ glucocorticoid production (see adrenal section)
Name the actions of GH that are mediated through IGF.	↑ protein synthesis in bone, muscle, and organs → ↑ linear growth, ↑ lean body mass, and ↑ organ size

Endocrine and Reproductive Physiology

What factors ↑ GH secretion?
1. GHRH
2. Sleep
3. Stress
4. Exercise
5. Starvation
6. Hypoglycemia
7. SS

What factors ↓ GH secretion?
1. GH and IGF (negative feedback)
2. Obesity
3. Hyperglycemia

Diagram the GH feedback loop.

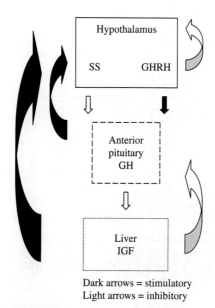

Dark arrows = stimulatory
Light arrows = inhibitory

Figure 7.5 GH Feedback Loop.

What factors ↑ PRL secretion?
1. Pregnancy
2. Breast-feeding
3. Stress
4. TRH
5. Dopamine antagonists

What factors ↓ PRL secretion?
1. PRL (negative feedback)
2. Dopamine (PIF)
3. Dopamine agonists (e.g., bromocriptine)
4. SS

Name the actions of the posterior pituitary hormones:

Oxytocin
1. ↑ contraction of myoepithelial cells in breasts (↑ milk ejection)
2. ↑ contraction of uterus

ADH
1. ↑ H_2O permeability of the distal tubule and collecting duct
2. Constricts vascular smooth muscle

What type of hormones are oxytocin and ADH?
Polypeptide hormones

Where is oxytocin synthesized?
Paraventricular nuclei of the hypothalamus

Where is ADH synthesized?
Supraoptic nuclei of the hypothalamus

Where are oxytocin and ADH stored and released?
Posterior pituitary

How are oxytocin and ADH synthesized and secreted?
Precursor
↓
Cleaved and packaged into secretory granules with neurophysins (carrier proteins)
↓
Transported by axoplasmic flow to posterior pituitary

How are the effects of ADH mediated, and what is its second messenger system?
1. Renal effect: V_2 receptor → cAMP
2. Smooth muscle effect: V_1 receptor IP_3

What factors regulate oxytocin secretion?
1. Breast-feeding
2. Sight or sound of infant
3. Dilation of cervix

What factors ↑ ADH secretion?
1. High serum osmolarity
2. Volume depletion
3. Pain
4. Nausea
5. Hypoglycemia
6. Nicotine
7. Opiates

What factors ↓ ADH secretion?
1. Low serum osmolarity
2. Atrial natriuretic peptide (ANP)
3. α-Agonists
4. Ethanol

What effect does lithium have on the body's response to ADH?
Decreases the response

ADRENAL GLAND

What are the three zones of the adrenal cortex (from outer to inner zones)?	1. Zona **G**lomerulosa 2. Zona **F**asciculata 3. Zona **R**eticularis *Remember: "**GFR**"
What are the special cells of the adrenal medulla called?	Chromaffin cells
What are the embryological origins of chromaffin cells?	Neural crest cells
Name the hormones of the:	
Adrenal cortex (from outer to inner zones)	Mineralocorticoids: aldosterone Glucocorticoids: cortisol Androgens: dehydroepiandrosterone (DHEA), androstenedione *Remember: "salt, sugar, and sex"
Adrenal medulla	Catecholamines: epinephrine (Epi), norepinephrine (NE)
From what are the adrenocortical (steroid) hormones derived?	Cholesterol

Diagram the steroid hormone pathway.

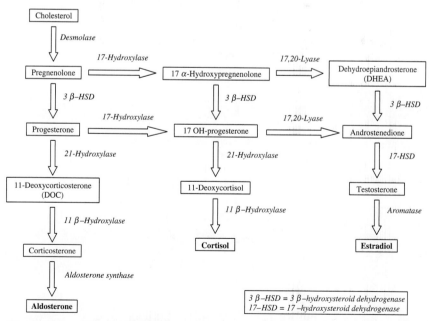

Figure 7.6 Steroid Pathway.

What is special about the enzymes of the steroid hormone pathway?	Most are members of the cytochrome P450 system
What is the rate-limiting step in the synthetic pathway?	Cholesterol desmolase
How is this step regulated?	ACTH
Name the actions of ACTH.	1. ↑ activation of desmolase 2. ↑ cortisol synthesis 3. ↑ lipolysis 4. ↑ cholesterol uptake into adrenal cortex 5. ↑ proliferation of zona fasciculata if ACTH is elevated for days
What factors ↑ ACTH secretion?	1. CRH 2. Circadian rhythm 3. Emotions/stress 4. Central nervous system (CNS) trauma
What factors ↓ ACTH secretion?	Cortisol (negative feedback)
Diagram the relationship between CRH, ACTH, and cortisol.	

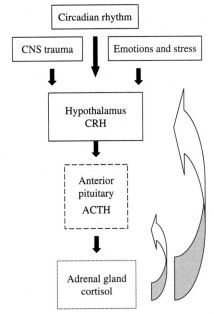

Dark arrows = stimulatory
Light arrows = inhibitory

Figure 7.7 CRH, ACTH, Cortisol Loop.

Endocrine and Reproductive Physiology

Name the actions of cortisol.	1. ↑ hepatic gluconeogenesis 2. ↓ protein synthesis 3. ↑ protein degradation 4. Facilitate interaction between Epi and NE and glucagon for normal metabolism 5. Facilitate interaction between PRL and insulin for certain growth effects 6. ↓ insulin 7. ↓ immune/inflammatory response 8. ↓ ACTH secretion (negative feedback)
What factors regulate cortisol synthesis?	ACTH
Name the actions of aldosterone.	1. ↑ Na^+ resorption in renal distal tubules 2. ↑ K^+ and H^+ excretion Net effect = ↑ volume
What factors regulate aldosterone synthesis?	1. Renin-angiotensin II-aldosterone system 2. ↑ K^+ 3. Some control by ACTH
Describe the renin-angiotensin II-aldosterone pathway.	Angiotensinogen ↓ Renin ← hypovolemia Angiotensin I ↓ Angiotensin converting enzyme (ACE) Angiotensin II ↓ Aldosterone release
When is the renin-angiotensin II-aldosterone system activated?	1. ↓ blood volume 2. ↓ serum Na^+
What cells sense hypovolemia and release renin in response?	Macula densa and juxtaglomerular cell, respectively
Where is ACE found?	Lungs (major) and vasculature (minor)
How are all these steroid hormones inactivated and excreted?	1. Catabolized by liver (majority) 2. Excreted through urine and bile/stool
From what are the catecholamines derived?	Tyrosine

Diagram the catecholamine pathway.

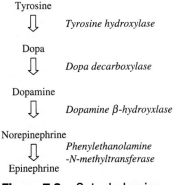

Figure 7.8 Catecholamine Pathway.

How are the effects of Epi and NE mediated, and what are their second messenger systems?	Via α and β adrenergic receptors: α_1 receptor → intracellular Ca^{2+} α_2 receptor → ↓ cAMP β_1 receptor → ↓ cAMP β_2 receptor → ↓ cAMP
Name the actions of Epi and NE.	
Metabolic	1. ↑ basal metabolic rate (BMR): ↑ O_2 consumption, ↑ heat production 2. ↑ glycogenolysis 3. ↑ lipolysis 4. ↑ glucagon secretion (β_2) → hyperglycemia
Eyes	Mydriasis (α_1)
Cardiovascular	1. ↑ Cardiac inotropy, chronotropy, dromotropy (β_1, β_2) 2. ↑ myocardial excitability → extrasystoles or arrhythmias 3. ↑ systolic and diastolic BP 4. Systemic vasoconstriction (α_1, α_2) 5. Skeletal, coronary vasodilation (β_2)
Pulmonary	1. Bronchodilation (β_2) 2. ↓ bronchial gland secretion (α_1)
Renal	↑ renin secretion (β_1)
GI	↓ motility and secretions (α and β)

What factors regulate Epi and NE synthesis?	1. Sympathetic stimulation 2. Stress 3. Trauma 4. Surgery 5. Exercise 6. Hypoglycemia 7. Nicotine
What are the half-lives of Epi and NE?	Approximately 2 minutes in circulation
How are catecholamines metabolized?	Nerve endings: Norepinephrine ↓ monoamine oxidase (MAO) Deaminated derivatives Liver: NE, Epi, deaminated derivatives ↓ catechol-*O*-methyltransferase (COMT) Metanephrines → urine ↓ Vanillylmandelic acid (VMA) → urine
What are the percentages of catecholamine derivatives found in urine?	~50% metanephrines ~35% VMA

TESTIS

What is genetic or chromosomal sex?	XY for male (♂) or XX for female (♀)
What is gonadal sex?	Presence of testes in ♂ or ovaries in ♀
What is phenotypic or somatic sex?	Characteristics of internal/external genitalia
What factor determines ♂ gonadal sex?	Testicular differentiation factor (TDF), produced by Y chromosome
What three factors determine ♂ phenotypic sex?	1. Müllerian inhibiting factor (MIF): inhibits Müllerian (female) duct from developing into uterus and tubes 2. Testosterone: stimulates wolffian (male) duct to differentiate to epididymis, seminal vesicles, and vas deferens 3. Dihydrotestosterone (DHT): stimulates urogenital sinus and tubercle to differentiate to penis, urethra, prostate, and scrotum

Why do the testes descend into the scrotum?	To maintain temperature ~2°C below core body temperature, which is vital for normal spermatogenesis
Name the main cell types of the testis and their associated product(s).	Sertoli cell: supportive environment for spermatogenesis, androgen-binding protein, MIF, Inhibin Leydig cell: testosterone Germ cell: spermatozoa (sperm)
Name the anatomic components of the testis and their associated function(s).	Seminiferous tubules (85% testis mass): spermatogenesis by Sertoli and germ cells Rete testis: connects tubules and efferent ductules Efferent ductules: transports sperm to epididymis by ciliary motion and contraction Epididymis: reservoir and site of further morphologic and functional changes to sperm Vas deferens: propels sperm into urethra by muscular contractions
Describe spermatogenesis.	Spermatogonium (primitive germ cell) ↓ 1° spermatocyte (occurs in adolescence) ↓ meiotic division 2° spermatocyte ↓ Spermatid ↓ Spermatozoa (mature sperm)
How long does it take for spermatogonia to mature into spermatozoa?	~74 days
What is the blood-testes barrier?	Tight junctions that protect spermatogenesis by preventing proteins from moving from the interstitium to the lumen of the seminiferous tubules
What factors regulate spermatogenesis?	1. GnRH 2. FSH 3. Inhibin (negative feedback) 4. Testosterone (paracrine)

How is testosterone synthesized?	Cholesterol
	   ↓ cholesterol desmolase
	Pregnenolone
	   ↓ 17 α-hydroxylase

Endocrine and Reproductive Physiology

17-hydroxy-pregnenolone
↓ 17, 20-lyase
DHEA
↓ 3 β-hydroxysteroid dehydrogenase
Androstenedione
↓↑ 21 β-hydroxylase
Testosterone

Name the actions of testosterone.
1. Embryonic differentiation of wolffian ducts to ♂ reproductive tract
2. Puberty
3. ♂ secondary sexual characteristics
4. Contribute to Sertoli cells' maintenance of spermatogenesis

What are the ♂ secondary sexual characteristics?
1. Growth of penis, epididymis, vas deferens, and prostate
2. Growth spurt
3. Voice changes
4. ↑ muscle mass
5. ↑ sex drive

What factors regulate testosterone secretion?
1. GnRH
2. LH
3. Testosterone (negative feedback)

Diagram the hypothalamus-pituitary-gonadal (HPG) axis of testosterone.

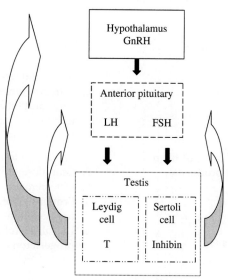

Dark arrows = stimulatory
Light arrows = inhibitory

Figure 7.9 HPG Axis for Testosterone.

Name the ♂ accessory genitalia and their associated functions.	Seminal vesicles: provides fructose to nourish sperm and secretes 60% of semen fluid content Prostate: secretes 20% of semen fluid content Bulbourethral (Cowper's) and urethral (Littre's) glands: add fluid during sperm transit through urethra
What is the relationship between testosterone and DHT?	Testosterone ↓ 5-α-reductase DHT
Where is 5-α-reductase found?	Prostate
Name the actions of DHT.	1. Embryonic differentiation of urogenital sinus/tubercle to ♂ genitalia 2. ♂ secondary sexual characteristics: male hair distribution 3. Growth of seminal vesicles
What factor stimulates puberty?	GnRH in pulsatile release

OVARY AND PLACENTA

What factor determines ♀ gonadal sex?	Absence of TDF: embryonic *indifferent gonads* automatically become ovaries
What two factors determine ♀ phenotypic sex?	1. Absence of MIF: Müllerian duct develops into uterus, fallopian tubes, and upper vagina 2. Estrogen: stimulates urogenital sinus and tubercle to differentiate into lower vagina, clitoris, and vulva
Name the main cell types of the ovary and their associated product(s).	1. Theca cell: testosterone 2. Granulosa cell: estradiol (from testosterone) 3. Primordial follicle: mature ovum for ovulation 4. Luteal cell (of corpus luteum): progesterone

What enzyme converts testosterone to estradiol?	Aromatase
Describe the menstrual cycle and oogenesis cycle:	
Follicular (proliferative) phase	Days 1–14 (variable): 1. Estradiol ↑ and progesterone ↓ 2. FSH and LH ↓ (negative feedback) 3. FSH and LH receptors ↑ (up-regulation) 4. Multiple primordial follicles enlarge (only one becomes the graafian follicle, others undergo atresia)
Ovulation	Day 15: 1. Estradiol ↑ ↓ Positive feedback ↓ LH↑↑ ↓ Ovulation! This phenomenon is known as the *estrogen-induced LH surge* 2. Temporary estradiol ↓ after ovulation
Luteal (secretory) phase	Days 15–28 (fixed): 1. Progesterone ↑ and estradiol ↑ 2. Corpus luteum matures and produces progesterone 3. Endometrium vascularity builds to prepare for implantation of fertilized ovum 4. Corpus luteum regresses if no fertilization occurs → progesterone ↓ and estradiol ↓

Diagram the menstrual and oogenesis cycle.

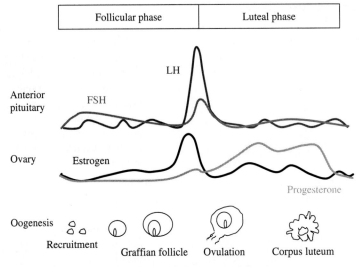

Figure 7.10 Menstrual and Oogenesis Cycle.

Why is the luteal phase fixed in duration?	Corpus luteum has a fixed 14-day life span
How long is the typical menstrual cycle?	Average of 28 days
When does menstruation occur?	Onset of menses marks Day 1 of menstrual cycle
What is menstruation?	Sloughing of endometrium due to ↓ progesterone and ↓ estradiol
How is estrogen synthesized?	Cholesterol ↓ cholesterol desmolase Pregnenolone ↓ ↓ ↓ Androstenedione ↓ 21-β-hydroxylase Testosterone ↓ aromatase Estradiol
Name the actions of estrogen.	1. Maturation of fallopian tubes, uterus, cervix, and vagina 2. Puberty 3. ♀ secondary sexual characteristics: growth of breasts 4. Development of granulosa cells

Endocrine and Reproductive Physiology

	5. Maintenance of pregnancy: suppression of uterine response to contractile stimuli and prolactin secretion
	6. Both positive and negative feedback on FSH and LH secretion
	7. Up-regulation of LH, estrogen, and progesterone receptors
Name the actions of progesterone.	1. Maintenance of pregnancy: suppression of uterine response to contractile stimuli
	2. Maintenance of luteal phase and uterine secretory activity
	3. Negative feedback on FSH and LH
What factors regulate the menstrual cycle, estrogen and progesterone secretion?	1. GnRH
	2. FSH and LH
	3. Estrogen (negative feedback)
	4. Estrogen (positive feedback—ovulation)
	5. Progesterone (negative feedback)

Diagram the HPG axis for estrogen and progesterone.

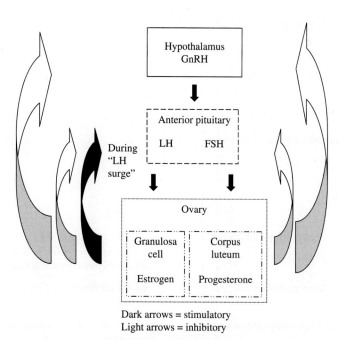

Dark arrows = stimulatory
Light arrows = inhibitory

Figure 7.11 HPG Axis for Estrogen and Progesterone.

Name the main hormone producers of pregnancy and their associated products.	Corpus luteum: 1. hCG 2. Estradiol 3. Progesterone (conception to ~week 12) Fetal adrenal gland: DHEA-S Placenta: 1. Estriol 2. Progesterone (week 6 to delivery) 3. HPL
How is estriol synthesized in pregnancy?	DHEA-S (fetal adrenals) ↓ aromatase Estriol (placenta)
How do the concentrations of the following hormones change during pregnancy:	
Pituitary:	
TSH	↔
LH and FSH	Basal level
GH	↔
PRL	↑ to term
ACTH	↔
Placenta and fetus:	
Estradiol	↑ to term
Estriol	↑ to term
Estrone	↑ to term
Progesterone	↑ to term
HPL	↑ to term
Maternal adrenals and ovaries:	
Testosterone	↑↑↑
DHEA	↓
Cortisol	↑
hCG	↑ then ↓
What factors influence lactation?	↑ of PRL through pregnancy accompanied by the ↓ of estrogens and progesterones after delivery
How is lactation maintained?	By breast-feeding, which ↑ PRL and oxytocin
How is ovulation suppressed during lactation?	PRL inhibits GnRH and therefore LH and FSH

PANCREAS

Name the pancreatic endocrine hormones.	1. Insulin 2. Glucagons 3. SS
What metabolic process do they influence?	Serum glucose regulation
Where is insulin synthesized?	β Cells of islets of Langerhans
What is insulin?	A polypeptide hormone
What are the steps of insulin synthesis?	Preproinsulin ↓ Proinsulin (C-peptide joins chains A and B) ↓ Insulin + C-peptide
What is the importance of C-peptide?	It distinguishes an endogenous source of insulin from an exogenous source
Where is insulin secreted?	Into portal circulation
How are the effects of insulin mediated?	Through insulin receptors on various tissues
What is its second messenger system?	Insulin receptor is a tyrosine kinase, which phosphorylates itself and other proteins
Name the actions of insulin in:	
Skeletal and cardiac muscle	1. ↑ glucose uptake via Glut4 (*insulin-sensitive* glucose transporter) 2. ↑ active transport of amino acids 3. ↑ protein synthesis 4. ↓ protein degradation 5. ↑ K^+ into cells
Liver	1. ↓ gluconeogenesis 2. ↓ glycogenolysis 3. ↑ glycogenesis
Adipocytes	↑ triglyceride synthesis, ↓ lipolysis
What is the overall effect of insulin on serum levels of:	
Glucose	Decrease
Amino acids	Decrease
Fatty acids	Decrease
Ketoacids	Decrease
K^+	Decrease

What factors regulate insulin secretion?	1. ↑ blood glucose 2. ↑ amino acids (especially leucine, arginine) 3. ↑ fatty acids 4. Gastric inhibitory peptide (GIP) 5. Vagus nerve stimulation 6. GH 7. SS (inhibitory effect)
How do the pancreatic β-cells *sense* serum glucose?	Via Glut2 (*glucose sensor* glucose transporter) Glucose binds Glut2 on β-cells ↓ Glucose metabolized to ATP ↓ β-Cell membrane depolarization ↓ Insulin release from pancreas
How does insulin secretion respond to blood glucose?	Biphasic response: 1. Rapid burst of insulin upon glucose exposure 2. Slowly rising release of insulin
Describe insulin receptor regulation in:	
Starvation state	Up-regulation
Obesity	Down-regulation
Where is glucagon synthesized?	α-Cells of islets of Langerhans
What is glucagon?	A single-chain polypeptide hormone
Where is glucagon secreted?	Into portal circulation
How are the effects of glucagon mediated?	Glucagon receptor
What is its second messenger system?	cAMP
Name the actions of glucagon on:	
Liver	1. ↑ glycogenolysis 2. ↑ gluconeogenesis 3. ↑ lipolysis 4. ↓ protein degradation 5. ↓ protein synthesis
Adipocytes	↑ lipolysis
What is the overall effect of glucagon on serum levels of:	
Glucose	Increase
Fatty acids	Increase
Ketoacids	Increase

What factors regulate glucagon secretion?	1. ↓ blood glucose 2. ↑ amino acids (especially arginine and leucine) 3. Epi 4. NE 5. Glucocorticoids 6. Cholecystokinin 7. SS (inhibitory)
Where is SS synthesized?	1. Delta cells of islets of Langerhans 2. Intestines 3. Nervous system
What is SS?	A polypeptide hormone—secreted in two forms
Name the actions of SS (pancreatic form).	1. ↓ insulin 2. ↓ glucagon 3. ↓ many other GI hormones

THYROID GLAND

What chemical element is required for normal thyroid function?	Iodine
Where are thyroid hormones synthesized?	Follicular cells of the thyroid gland
Name the two active thyroid hormones.	T_4 (thyroxine)—most produced T_3 (3, 5, 3'-triiodothyronine)—most active
What protein synthesizes and stores thyroid hormones?	Thyroglobulin—contains the iodotyrosyl residues monoiodotyrosine (MIT) and diiodotyrosine (DIT)
What are the steps of thyroid hormone synthesis?	1. Inorganic iodide (I^-) actively transported into follicular cell → diffuse into colloid/lumen → oxidized to iodine (I_2) via thyroid peroxidase 2. Thyroglobulin produced in rough ER of follicular cell → secreted into colloid 3. In colloid, I_2 is incorporated into tyrosine residues of thyroglobulin → MIT and DIT formed 4. Coupling of iodotyrosyl residues inside thyroglobulin occur to form T_3 (MIT + DIT) or T_4 (DIT + DIT)

Diagram thyroid hormone synthesis.

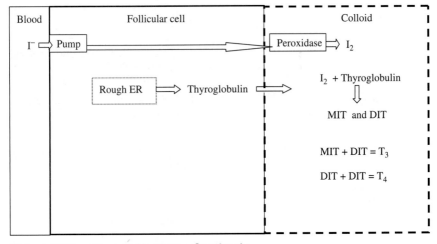

Figure 7.12 Thyroid Hormone Synthesis.

What are the steps of thyroid hormone secretion?	1. Thyroglobulin is resorbed into follicular cell → fuse with lysosomes hydrolyzed into T_3, T_4, DIT, MIT 2. T_3 and T_4 secreted into plasma 3. DIT and MIT deiodinated → tyrosines
Where else are thyroid hormones produced?	80% of T_3 is converted from T_4 in peripheral tissues, including liver and kidney
What else does T_4 form peripherally?	T_4 → deiodinated → inactive rT_3
How do thyroid hormones circulate?	1. >99% in circulation are bound to proteins (e.g., thyroxine-binding globulins [TBG], albumin, and prealbumin) 2. Freely circulating T_3 and T_4 are active
What are the half-lives of T_3 and T_4?	T_3: 1 day T_4: 7 days

Name the actions of T_3 and T_4 on:	
Metabolism	1. ↑ BMR: ↑ O_2 consumption, ↑ heat production, activate Na^+-K^+-ATPase 2. ↑ glucose absorption, glycogenolysis, gluconeogenesis, glucose oxidation 3. ↑ lipolysis and protein degradation
Growth/development	1. Required for actions of GH to promote linear growth/bone formation 2. Required for fetal CNS development
Cardiac	1. ↑ β-adrenergic receptors → ↑ cardiac output (HR × SV) 2. ↑ systolic blood pressure only
Respiratory	↑ ventilation rate
Reproductive	Required for ovary and testis maturation
What factors regulate thyroid hormone synthesis and secretion?	1. TSH 2. Thyrotropin releasing hormone (TRH)
What is TSH?	A glycoprotein with α- and β-subunit
Where is TSH synthesized?	Anterior pituitary
How are the effects of TSH mediated?	Through TSH receptor on follicular cells
What is its second messenger system?	cAMP
Name the actions of TSH.	↑ all aspects of thyroid hormone synthesis and secretion
What is TRH?	A polypeptide hormone
Where is TRH synthesized?	Hypothalamus
How are the effects of TRH mediated?	Through TRH receptor on pituitary cells
What is its second messenger system?	IP_3
Name the actions of TRH.	↑ pituitary secretion of TSH

Diagram the TRH, TSH, and thyroid hormone loop.

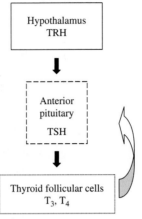

Dark arrows = stimulatory
Light arrows = inhibitory

Figure 7.13 TRH, TSH, and Thyroid Hormone Loop.

PARATHYROID GLAND

Name the physiological processes that involve Ca^{2+}.	1. Muscular contraction 2. Membrane permeability 3. Endocrine and exocrine secretions 4. Enzyme regulation 5. Coagulation
How does Ca^{2+} circulate in serum?	1. 40–50% in free, ionized form (biologically active) 2. 50–60% bound to plasma proteins or complexed to other ions
Where is Ca^{2+} stored and approximately in what amounts?	1. ECF: 0.9 g 2. ICF: 11 g 3. Bone: 1000 g
How is Ca^{2+} stored in bone?	1. Two-thirds as inorganic crystals = hydroxyapatite $[Ca_{10}(PO_4)_6(OH)_2]$ 2. One-third as organic materials (e.g., Ca_2PO_4)
Name the three organs that play an important role in serum Ca^{2+} regulation.	1. Intestines → absorption 2. Kidney → excretion 3. Bone → resorption and formation
How is Ca^{2+} homeostasis maintained?	Net absorption must be balanced by excretion

Diagram Ca^{2+} metabolism.

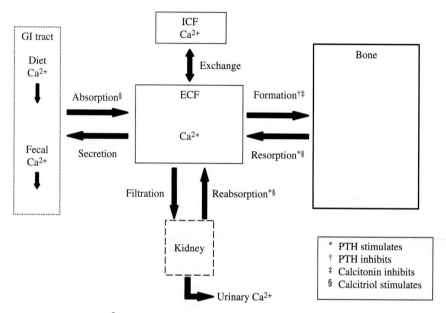

Figure 7.14 Ca^{2+} Metabolism.

Name the three hormones that play an important role in Ca^{2+} homeostasis.	1. Parathyroid hormone (PTH) 2. Calcitriol (1, 25-dihydroxyvitamin D) 3. Calcitonin
What is PTH?	A polypeptide hormone containing 84 amino acids, only the first 34 amino acids are required for biological activity
Where is PTH synthesized and secreted?	Chief cells of the four parathyroid glands
How is PTH synthesized?	Pre-proPTH → proPTH → PTH
What factor regulates PTH secretion?	Serum Ca^{2+} (negative feedback)
What happens to PTH secretion when	
Serum Ca^{2+} is low	Increase
Serum Ca^{2+} is high	Decrease
How are the effects of PTH mediated?	Through the PTH receptor
What is its second messenger system?	cAMP and intracellular Ca^{2+}

Name the actions of PTH on the following:

Kidney
1. ↑ Ca^{2+} reabsorption → ↑ serum Ca^{2+} (distal tubule)
2. ↓ PO_4^{4-} reabsorption → ↑ serum Ca^{2+} (proximal tubule)
3. ↑ 1-α hydroxylase → ↑ calcitriol production → ↑ serum Ca^{2+} (proximal tubule)
4. Stimulates osteoclasts to increase bone resorption → ↑ Ca_2PO_4 in ECF

Bone
Inhibits osteoblasts to further increase bone resorption

What is vitamin D (cholecalciferol)?
A steroid hormone synthesized by skin through a UV radiation reaction

What is the active form of vitamin D?
Calcitriol (1, 25-dihydroxycholecalciferol)

What are the inactive forms of vitamin D?
1. Cholecalciferol
2. 25-Hydroxycholecalciferol
3. 24, 25-Hydroxycholecalciferol

Where and how is calcitriol synthesized?
Vitamin D
↓ 25-Hydroxylase (in liver)
25-Hydroxycholecalciferol
↓ 1 α-Hydroxylase (in kidney)
1, 25-Hydroxycholecalciferol and 24, 25-Hydroxycholecalciferol

What factors regulate calcitriol synthesis?
Activity of 1 α-hydroxylase, which is ↑ by
↓ serum Ca^{2+}
↑ PTH levels
↓ serum PO_4^{4-}

How are the effects of calcitriol mediated?
Through the calcitriol receptor

Name the actions of calcitriol on:

Intestines
1. Induces Ca^{2+}-binding protein → ↑ Ca^{2+} absorption
2. ↑ PO_4^{4-} reabsorption

Kidney
↑ Ca^{2+} and PO_4^{4-} reabsorption

Bone
↑ bone resorption → ↑ serum Ca_2 and PO_4^{4-}

What is calcitonin?
A polypeptide hormone of 32 amino acids

Where is calcitonin synthesized and secreted?
Parafollicular cells (C cells) of thyroid gland

What factor regulates calcitonin secretion?	↑ serum Ca^{2+} stimulates its secretion
What is its second messenger system?	cAMP (most likely)
Name the actions of calcitonin.	1. ↓ bone resorption by osteoclasts 2. ↓ bone formation by osteoblasts → ↑ serum Ca^{2+}

Summarize the factors that regulate Ca^{2+} homeostasis hormone secretion:

PTH	↓ serum Ca^{2+}
Calcitriol	↓ serum Ca^{2+} ↑ PTH levels ↓ serum PO^{4-}
Calcitonin	↑ serum Ca^{2+}

Summarize the actions of the Ca^{2+} homeostasis hormones at each site:

Intestines	PTH: ↑ Ca^{2+} absorption through activating calcitriol Calcitriol: ↑ Ca^{2+} and PO^{4-} reabsorption Calcitonin: No effect
Kidney	PTH: ↑ Ca^{2+} reabsorption and ↓ PO^{4-} reabsorption. Calcitriol: ↑ Ca^{2+} and PO^{4-} reabsorption Calcitonin: No effect
Bone	PTH: ↑ resorption Calcitriol: ↑ resorption Calcitonin: ↓ resorption

Summarize the net effect of the Ca^{2+} homeostasis hormones on serum Ca^{2+} and PO^{4-}:

	Serum Ca^{2+}	Serum PO^{4-}
PTH	Increase	Decrease
Calcitriol	Increase	Increase
Calcitonin	Decrease	

CHAPTER 8
Clinical Vignettes

A young man is inadvertently given a dose of tetrodotoxin and his excitable cells are no longer able to generate action potentials, how did this occur?	Voltage-sensitive Na^+ channels are blocked and action potentials cannot be produced
What abnormal state is associated with accommodation, high serum K^+ concentrations, and muscle weakness?	Hyperkalemia
Some of a patient's blood is collected and an unknown solution is added to the sample. When the technician views the slide, all of the red blood cells (RBCs) have lysed. What is the osmolarity of the solution relative to the patient's sample?	Hypotonic
A drug company is working on a drug that has maximum efficacy intracellularly. What attribute of their drug must they increase to improve permeability through the lipid bilayer cell membrane?	Oil/water partition coefficient
A young woman is poisoned by a jealous friend with an agent that inhibits her cell's Na^+-K^+-ATPase. What will happen to the concentrations of Na^+, K^+, and Ca^{2+}?	1. Increase in intracellular Na^+ 2. Decrease in intracellular K^+ 3. Increase in intracellular Ca^{2+} from decreased Na^+ gradient
A researcher is performing experiments on the visual system in some test animals. What will be the effect on the animal's vision if transections in the following locations are made:	
Single optic nerve	Ipsilateral blindness
Optic chiasm	Heteronymous bitemporal hemianopia
Optic tract	Homonymous contralateral hemianopia
Geniculocalcarine tract	Homonymous hemianopia with macular sparing

A researcher is studying muscular contraction and has set up an experiment whereby some muscle fibers are constantly stimulated. Once he begins the stimulation he notes that the muscles enter a state of tetanic contraction. What ion is responsible for this and what would its concentration be inside the muscle cells?	Ca^{2+} is responsible for the tetany. It would be at very high levels due to constant release by the sarcoplasmic reticulum (SR) from the constant stimulation the muscle fibers are receiving and also because it is not being reaccumulated
Drug X is an inhibitor of acetylcholinesterase (AChE) for which AChE has a very high affinity. What would be the effect on the potential generated if drug X were given to someone?	Degradation of acetylcholine (ACh) would be blocked and its action would be prolonged and produce a larger end-plate potential
Detectives get a call about a shooting. When they arrive on scene, they find a young man has been shot and killed. As they are inspecting the body for clues, it is noted to be in a rigid state. What is the physiologic basis for this state?	Rigor mortis is due to the lack of adenosine triphosphate (ATP). After a person dies, they are no longer able to regenerate ATP, thus myosin and actin in skeletal muscle fibers remain tightly bound, producing a rigid state.
A scientist is investigating a substance given to him by a colleague. He applies the substance to an autonomic ganglia and sees that action potential propagation is blocked. When he applies it to a neuromuscular junction, there is no change in action potential generation. What is the substance likely to be?	Hexamethonium, which is a nicotinic antagonist at the ganglion, but does not have an effect on the neuromuscular junction
A young woman met with an accident in which her left thalamic nucleus is destroyed. What is the physiologic consequence of this lesion?	She will have no sensation on the right side of her body
A previously healthy, 37-year-old man presents to his primary care physician (PCP) complaining of headaches that seem to be increasing in frequency. When further questioned, he notes that he has also been having blurry vision. A brain magnetic resouance imaging (MRI) reveals a 1.6 cm soft-tissue mass on the sella turcica with suprasella extension. What structure has this tumor compressed to produce the visual symptoms?	The optic chiasm

Clinical Vignettes

A 45-year-old man presents with new-onset vision deficit in the right eye. On examination, when the left eye is illuminated, both pupils constrict; while on illumination of the right eye, both pupils dilate. When the light is moved from the left eye to the right eye, the right eye dilates. Where is the lesion?	The right retina or optic nerve
A group of cultists are brainwashed to avoid vitamin A as the "A" stands for anarchy. What visual deficiency will these individuals develop from this?	Night blindness, which is caused by vitamin A deficiency
A 6-year-old girl is spinning in the rightward direction on the playground with her friend when they both stopped and fell. What is the first event that occurs in her vestibular system?	The cupula in the left horizontal semicircular canal moves toward the utricle and bends the stereocilia toward the kinocilium, causing the hairs cells to depolarize
A 64-year-old woman complains of persistent nonproductive cough and weight loss of 7–10 lb in the past 3 months. On chest x-ray (CXR) and computed tomography (CT), a dense lesion in the right apex was visualized, which is later confirmed to be a carcinoma. On a subsequent follow-up visit, she reports deficiencies in her vision and her face is not sweating, even in Arizona's 90°F weather. You notice that her eyelids are drooping more so than previously. What do you suspect these symptoms mean?	The patient probably has a Pancoast tumor, which is compressing her sympathetic cervical ganglion causing Horner syndrome: miosis, ptosis, and anhydrosis
A pugilist is struck in the nose and breaks his cribriform plate. After he recovers from his fight, he notices that his wife's perfume does not smell as strongly as it usually does. What is the likely mechanism for this change?	The broken cribriform has likely severed some of the input fibers to the olfactory bulb and produced hyposmia
A high school boy is fond of listening to rock music that favors heavy bass (low-frequency) tones. He also has a tendency to listen to music at very high volumes. Which part of the basilar membranes in his ear is at risk for serious damage?	The apices, which detect low frequencies

A young child accidentally touches a pot filled with boiling water. He quickly withdraws his hand from the pain, but simultaneously extends his other hand to maintain his balance. What caused the extension of the other hand?	Crossed extension reflex, which is part of the flexor withdrawal reflex and helps to maintain balance
A young man is in a car accident in which his spinal cord is injured. When he is rushed into the trauma bay, his legs are flaccid and reflexes cannot be obtained. The patient overhears these results and is very frightened. What can the doctor tell him about the condition of his legs?	The patient is in spinal shock, whereby he has lost excitatory stimulation from both α- and γ-motoneurons. With time he may have partial recovery of muscle control and return of his reflexes with the possibility of hyperreflexia
A 50-year-old man slipped while he was shoveling his driveway and hit his head. His wife brings him into the ER. When the doctor goes to see him, it is noted that the man is unable to speak or write, but can understand commands. What type of injury has he suffered?	The man has a motor aphasia, which results from damage to Broca's area
A young woman is in the recovery room after undergoing an operation. She is still only partially awake, but her nurse notices that her skin is hot to touch and that her oxygen saturation has decreased. A diagnosis of malignant hyperthermia is made and appropriate steps are taken. What agent in her previous procedure likely caused this state?	An inhaled anesthetic was likely used. One of their major complications is induction of malignant hyperthermia
A patient has a mean arterial pressure (MAP) of 70 mm Hg, a right atrial pressure (RAP) of 10 mm Hg, and total peripheral resistance (TPR) is determined to be 10 mm Hg min/L. What is the cardiac output (CO)?	$CO = \dfrac{MAP - RAP}{TPR}$ Substituting in the numbers: $CO = \dfrac{70 \text{ mm Hg} - 10 \text{ mm Hg}}{10 \text{ mm Hg min}/L} = 6 \text{ L/min}$
An elderly gentleman has bilateral carotid bruits that are audible with auscultation. What is the physiologic mechanism for the increased turbulence that creates this phenomenon?	Increased blood velocity due to the narrowed vessel increases the turbulence
What happens to the stressed volume in older patients compared to younger ones?	It decreases since the capacitance of the arteries decreases with age

Clinical Vignettes

A 14-year-old boy is brought to the ER from his hockey game in which he was struck in the chest with the puck from a slap shot. He is unresponsive and a stat electrocardiogram (ECG) shows P waves and QRS complexes with no association. What type of heart block is the patient in?	3° (third degree) or complete heart block
A 50-year-old woman comes in for a preoperative examination and has an ECG done that shows no P wave, but a normal QRS complex and T wave. Where is the pacemaker of her heart located?	Atrioventricular (AV) node
An electrician is working with a live wire when he is suddenly shocked. Just immediately preceding the shock his ventricles contracted. Why do they not contract again from the shock?	His ventricles are in the absolute refractory period
A patient is given an experimental drug, which the manufacturer claims to be a cardioselective ACh-analog. What type of effects will this drug have on the patient's heart rate and conduction velocities?	The drug will mimic parasympathetic stimulation and result in 1. Decreased heart rate 2. Decreased conduction through AV node 3. Increased PR interval
An innocent bystander is hit by a stray bullet in a drive-by shooting and begins to bleed profusely. What is the response by his autonomic nervous system to the acute blood loss?	1. Increased sympathetic stimulation of his heart and vasculature 2. Decreased parasympathetic stimulation of his heart
A baby is born with a ventricular septal defect. What can be said about the blood flow out of the heart, if the left ventricular pressure is greater than that of the right?	Pulmonary blood flow is greater than aortic blood flow from the left-to-right shunt that is present
A 65-year-old man undergoes angioplasty of his left anterior descending coronary artery. After the procedure, the vessel's radius has doubled. What will be the change in resistance?	It will decrease by a factor of 16 *Remember: $$R = \frac{8\eta l}{\pi r^4}$$
A young woman who suffered severe total body burns is brought into the hospital with grossly edematous limbs. What is the physiologic cause for the edema?	Increased permeability of the capillaries to water (e.g., ↑ K_f) from damage suffered by the burns

What is the direction of fluid movement and net driving pressure in a capillary where P_c is 32 mm Hg, P_i is 3 mm Hg, π_c is 26 mm Hg, and π_i is 3 mm Hg?	Net pressure = $(P_c - P_i) - (\pi_c - \pi_i)$ = $(32) - (-3) - (26 - 3)$ = +12 mm Hg Fluid will move out of the capillary since the net pressure is positive
A patient experiences orthostatic hypotension after being started on a new medication. What is being impaired by the medication to cause this condition?	Carotid sinus baroreceptor response
In a patient with a stroke, what is the stimulus that causes the body's response to preserve blood flow to the brain?	P_{CO_2} in brain tissue, which increases with cerebral ischemia
A medical student is performing a cardiovascular examination on a 5-year-old patient and hears a third heart sound. Should he be concerned about this finding?	No. A third heart sound is considered normal in children, but is pathological in adults
A middle-aged woman comes to the ER complaining of chest discomfort and a feeling of light-headedness. On examination, she has an irregularly irregular pulse. What abnormal rhythm is this patient likely experiencing?	Atrial fibrillation
A young man is stabbed in the chest during an altercation and begins to bleed into his pericardium. What happens to the cardiac output and pressure in his heart chambers from this situation?	Cardiac output: decreases Pressure: equalizes in all four chambers
Which obstructive lung disease causes a reduction in diffusing capacity of the lung for carbon monoxide (DLCO)?	Emphysema
A 65-year-old male with a 75-pack-year smoking history presents to the pulmonologist for evaluation of his dyspnea and exercise intolerance. What values does the pulmonologist expect to see for his pulmonary function tests (PFTs)?	1. ↓ forced expiratory volume (FEV) 2. ↓ FEV_1/forced vital capacity (FVC) 3. ↑ total lung capacity (TLC)
A 45-year-old female with severe scoliosis and no history of tobacco use complains of dyspnea. What category of lung disease does she have?	Restrictive lung disease

A 40-minute-old male neonate of an 18-year-old Caucasian female at 27 weeks gestation is noted to have central cyanosis. His respirations are shallow and rapid at 65/min. Other vital signs are stable. On examination, there is nasal flaring, audible grunting, and duskiness with intercostals and subcostal retractions. Fine rales are heard over both lung bases. Nasal O_2 does not improve his cyanosis. A CXR reveals fine reticular granularity, predominantly in lower lobes. Arterial blood gas (ABG) reveals hypoxemia with metabolic acidosis. What is the underlying cause of his condition?	Decreased production and secretion of surfactant resulting in atelectasis and shunting (perfused but nonventilated alveoli)
How do the PFTs in a patient with scoliosis differ from those with parenchymal disease and associated restrictive lung disease?	DLCO is normal in a patient with scoliosis, whereas it is reduced in a patient with parenchymal disease
A 34-year-old male presents to his PCP with cough, rhinitis, and wheezing. The patient had missed his appointment for similar symptoms 3 weeks ago, stating he felt better after spending his 3-day weekend resting. The patient has been working in a textile dye factory for the past 9 months. He presents today with reportedly worsening cough and difficulty breathing over the last week. He almost cancelled today's appointment because his symptoms seemed better having rested over the weekend. What is the probable diagnosis based on this patient's history?	Occupational asthma
What test can be used to aid in the diagnosis of the above patient?	Measure peak expiratory flow (PEF) and correlate changes with workplace exposure
What class of antihypertensive drugs should patients with asthma avoid?	Nonselective beta-blockers (e.g., atenolol)

A 67-year-old woman (height: 65 in., weight: 110 lb) with a known history of systemic lupus erythematosus (SLE) presents with progressively nonproductive cough and increasing dyspnea on exertion. She denies weight loss, fever, and hemoptysis. On pulmonary examination: RR 26 and faint bibasilar rales are appreciated. PFTs: FVC 3.1 (59%), FEV_1 2.2 (56%), maximum voluntary ventilation (MVV) 90 L/min, DLCO 14 mL/min/mm Hg (44%). A CXR reveals increased linear markings at the bases bilaterally. Based on the patient's history and PFTs, what property of the lungs is most likely affected?	Compliance is decreased from her restrictive lung disease
In the patient above, an ABG at room air was obtained as follows: pH 7.44, Pa_{O_2} 60 mm Hg, and Pa_{CO_2} 35 mm Hg. What is the alveolar to arterial oxygen gradient?	A-a gradient = $F_iO_2(P_{atm} - P_{water\ vapor})$ $- (P_aO_2 + P_aCO_2/R)$ A-a = $21\%(760 - 47) - (60 + 35/0.8)$ A-a = 45.9
A 51-year-old man with known interstitial lung disease with resting ABG: pH 7.45, Pa_{O_2} 60 mm Hg, and Pa_{CO_2} 30 mm Hg. After walking on a treadmill for 5 minutes, another ABG was drawn with Pa_{CO_2} 30 mm Hg. What is the most likely mechanism for the worsening hypoxemia observed postexercise?	Abnormalities in diffusion. In a patient with interstitial lung disease where lung parenchyma is affected, diffusion becomes important during exercise
A 66-year-old woman with chronic obstructive pulmonary disease (COPD) presents to the ED with progressively worsening dypsnea and a chronic cough. On physical examination, the patient was noted to have bibasilar rhonchi, diffuse wheezing bilaterally, and decreased breath sounds. A CXR reveals large hilar shadows suspicious for large pulmonary arteries and right atrial enlargement, which was also seen on ECG. What does this patient likely have?	Pulmonary hypertension secondary to chronic COPD
A young man is in a motor vehicle accident in which he suffers trauma leading to the paralysis of his diaphragm. What changes will occur in his blood gases from his injury?	It causes global hypoventilation: 1. pH decreased 2. Pa_{O_2} decreased 3. Pa_{CO_2} increased 4. A-a gradient is unchanged

Clinical Vignettes

What is the most common cause of hypoxemia encountered in the clinical setting?	Ventilation-perfusion ($\dot{V}/\dot{Q}$) mismatch
An elderly patient, hospitalized for knee replacement surgery, was found on postop day 5 with a 3-day fever, productive cough, and upper airway congestion. A CXR reveals localized infiltrates in the right middle lobe suggestive of pneumonia. What would the resting ABG look like if one were to be obtained?	The patient is expected to have shunting due to probable pneumonia: 1. pH increased 2. Pa_{O_2} decreased 3. Pa_{CO_2} decreased 4. A-a gradient is widened Hypoxemia does not respond to increased FiO_2
A 22-year-old tall healthy male presents with sudden new-onset sharp chest pain and dyspnea. He denies any recent fevers or coughs. He also denies trauma to the chest. A CXR reveals a moderate-sized pneumothorax at the left apex and no other abnormalities. What happened to the patient's lung and chest wall in this condition?	The lung collapsed inward and the chest wall sprang outward
How does carbon monoxide poisoning affect arterial O_2 concentration?	Decreases it
A 67-year-old male with a 100-pack-year history of smoking presents to the ED with progressively worsening dyspnea. On examination, the patient appears barrel-chested. He is visibly gasping for air with pursing lips and using his accessory muscles. Lung examination reveals occasional wheezing and coarse stridorous breath sounds. Pulse oximetry shows O_2 sat of 86%. The patient is given a breathing treatment as well as oxygen therapy. Why must oxygen therapy be administered with caution in this patient who is in severe respiratory distress?	This patient most likely has severe COPD from his extensive smoking history. There is a chance he is also hypercapnic, in which case, the patient may be dependent on his hypoxic drive (via the carotid and aortic chemoreceptors) to stimulate respiration rather than CO_2 level. If the hypoxic drive is withdrawn by administering O_2, the patient may become apneic, causing the arterial Po_2 to drop and increase Pco_2. Breathing may not restart because the increase in Pco_2 will further depress his respiratory center
How does anemia affect Po_2?	Decreases mixed-venous Po_2
How may mixed-venous Po_2 be increased?	1. Increase O_2 delivery 2. Increase O_2 content of inspired air 3. Increase myocardial work (e.g., increase stroke volume [SV]) 4. Increase intravascular volume

Using the following diagram, describe how the body's water composition changes in the give situations:

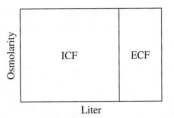

Figure 8.1 Fluid Compartments.

(This question shows up all over the boards, get familiar with them now.)

A young woman comes into the office after just returning from a trip to Mexico for her spring break. On her last day there she ate some food from a roadside vendor and developed diarrhea.	Diarrhea: extracellular fluid (ECF) decreases with no change in osmolarity. There is no change in the volume of the intracellular fluid (ICF)

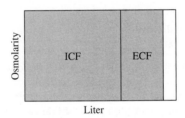

Figure 8.2 Diarrhea.

What changes will you see in the fluid distribution?

Some eager medical students want to prepare for their clinical rotations and decide to practice giving injections of isotonic saline to a willing volunteer. What happens to the fluid compartments as they do this?	Infusion of isotonic saline: the osmolarity of the fluid is the same compared to that of the body fluids, so it mostly stays in the ECF. Thus the volume of the ECF increases with no change in the ICF

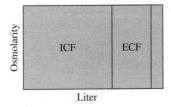

Figure 8.3 Isotonic Saline.

You go to your friend's house for the biggest sporting event of the year. He has all sorts of salty party food and you dig right in. Unfortunately the water to his apartment has been turned off, he's got nothing to drink in his fridge, and no one is willing to leave the game to get some drinks. What is happening to your fluid compartments?

Excessive NaCl intake: this increases the osmolarity of the ECF, thus drawing the fluid out of the ICF into the ECF. The result is a decrease in volume of the ICF and increase in volume of the ECF, while increasing the osmolarity of both ICF and ECF

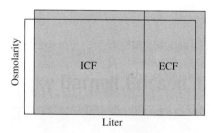

Figure 8.4 Excess NaCl.

A young lady has been recently diagnosed with adrenal insufficiency. Please draw the diagram and describe the situation.

Adrenal insufficiency: results in a loss of NaCl, thus ECF osmolarity decreases and the volume of the ECF also decreases. As a result, water diffuses into the ICF until the osmolarity of ICF and ECF are equal, raising the volume of the ICF and decreasing of the ECF

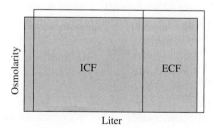

Figure 8.5 Adrenal Insufficiency.

A young woman who has recently been diagnosed with oat cell cancer of the lung comes into the office because she is feeling drowsy. Physical examination shows a hypertensive lady with dry mucous membranes, poor skin turgor, peripheral edema, and crackles in the base of the lungs. Routine labs show sodium of 123. A diagnosis of syndrome of inappropriate antidiuretic hormone (SIADH) is made. What is the fluid versus osmolarity body composition of a patient with SIADH? Describe why it is like this.

SIADH: this causes retention of free water by the kidneys, which results a net decrease of osmolarity of the both ICF and ECF. Their volumes both increase equally

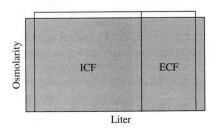

Figure 8.6 SIADH.

A second year medical student is studying for his step 1 United States Medical Licensing Examination (USMLE) when the air conditioner breaks in his apartment. It is a hot spring day and the student begins to sweat. It is Sunday and the library is closed, so the diligent student stays in his hot apartment and studies all day despite the heat. He is so bent on doing well in the examination that he forgoes drinking any fluids to maximize his study time. Describe the change of his body fluid dynamics and why it occurs.

Sweating: Sweat has more water than salt, so there is a net hyperosmotic volume contraction. ECF volume decreases, water shifts out of ICF leading to increased ICF osmolarity until it is equal to ECF osmolarity. As a result of all of this, there is still a net ICF and ECF volume loss with increased osmolarity of ICF and ECF

In what syndrome is the charge barrier lost and what are the consequences?

Nephrotic syndrome with resultant albuminuria, hypoproteinemia, generalized edema, and hyperlipidemia

Clinical Vignettes

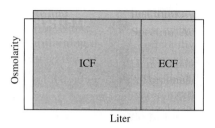

Figure 8.7 Sweating.

A man has been on trip to Mexico. After he returns, he goes to his doctor with severe diarrhea. An ABG shows the following values:
 pH = 7.2
 P_{CO_2} = 25
 $[HCO_3^-]$ = 8 mEq/L
Further lab values find Na^+ to be 135 and Cl^- to be 118. What is the anion gap? What acid-base disorder does this patient suffer from?

Anion gap = $[Na^+] - ([Cl^-] + [HCO_3^-])$
 = 135 − (115 + 8) = 12
This is a normal anion gap. With a low pH, we know that it is either metabolic or respiratory acidosis. Since the P_{CO_2} is normal and the HCO_3^- is low, it is a metabolic acidosis likely secondary to his diarrhea

A woman, complaining of frequent urination and occasional headaches, visits her doctor for further workup. Her plasma osmolarity is found to be 270 mOsm/L and urine osmolarity 1200 mOsm/L. What is the diagnosis for this patient?

SIADH. The plasma osmolarity is lower than normal, yet the urine osmolarity is still high. This is not a normal physiologic response and is indicative of SIADH

A volunteer is injected with paraaminohippuric acid (PAH) to measure renal plasma flow (RPF). Plasma concentration of PAH is 2 mg/mL, urine concentration of PAH is 500 mg/mL, and urine flow rate is 1 mL/min. What is the RPF?

Using the equation:
RPF = C_{PAH} = $([U]_{PAH} \times V)/[P]_{PAH}$
RPF = (500 mg/mL × 1 mL/min)/2 mg/mL = 250 mL/min

In the above patient, the hematocrit is found to be 40%. What is the renal blood flow? What is an approximation of the cardiac output in this patient?

Renal blood flow (RBF) =
 RPF/(1-hematocrit)
RBF = 250/(1 − 0.4) = 250/0.6
 = 416.67 mL/min
RBF is ~20% of cardiac output
So, RBF = CO × 0.2 or
CO = RBF/0.2 = 2083.4 mL/min

A middle-aged man comes to the ER complaining of 10/10 stabbing left-sided flank pain that started early in the morning. He does not report any fever or chills. On ultrasound, he is found to have a ureteral stone. What is the effect of this ureteral stone on glomerular filtration rate (GFR), RPF, and filtration fraction?	GFR will decrease because of the increased pressure in Bowman's space. There will be no change in RPF because there is no change in the blood flow to or from the kidneys. The filtration fraction will decrease because GFR has decreased and RPF has remained unchanged
A recently discovered drug (drug X) has been found to have a side effect of causing constriction of afferent renal arterioles. What will be the effect on the filtration fraction?	It will be unchanged! Even though there will be a decrease in RPF, there will also be a decrease in GFR from drop in P_{GC}
A young woman comes to the doctor after being treated for a bad urinary tract infection. Her GFR is found to be 100 mL/min. Her plasma glucose level is found to be 150 mg/dL and urine glucose concentration level is found to be 1000 mg/dL with flow at 3 mL/min. Determine whether there is net absorption or secretion of glucose.	Rate filtered = GFR × Plasma [glucose] = 100 mL/min × 150 mg/dL = 150 mg/min Rate secreted = V × urine [glucose] = 3 mL/min × 1000 mg/dL = 30 mg/min Difference = 150 − 30 = +120 mg/min Therefore, the glucose is reabsorbed
What commonly used cardiac drug can cause hyperkalemia (high K^+)? How?	Digitalis—blocks the Na^+-K^+ pump
What happens to the urinary secretion of Ca^{2+} in a person on loop diuretics (e.g., furosemide)?	Increases (loop diuretics inhibit Na^+ reabsorption, which is coupled to Ca^{2+} reabsorption)
When someone drinks excess water, what happens to the osmoreceptors in the anterior hypothalamus?	Inhibited
What is a condition in which antidiuretic hormone (ADH) is ineffective?	Nephrogenic diabetes insipidus
What medication may cause ADH to become ineffective?	Lithium
A 50-year-old woman presents to her doctor with tinnitus. During examination, she was found to have rapid breathing, and an ABG showed a P_{CO_2} of 24. Her labs were found to be:	1. Calculate serum anion gap: $[Na^+] − ([Cl^-] + [HCO_3^-]) = 140 − (105 + 15) = 20$ From this, we know that she has a metabolic acidosis, with an anion gap.

Na⁺ = 140
K⁺ = 5
Cl⁻ = 105
HCO₃⁻ = 15
BUN = 43
Creatinine = 1.7

What are the acid-base disturbances? What drug(s) can cause these types of disturbances?

2. Calculate proper response to the acidosis. For every mEq/L drop in [HCO₃⁻], there should be a compensatory drop of 1.3 mm Hg of P_{CO_2}. Normal [HCO₃⁻] is 24, so there is a drop of 7 mEq/L, so there should be a drop of 9.1 mm P_{CO_2} (normal value P_{CO_2} is ~40 mm Hg). Therefore, P_{CO_2} should be 40 − 9 = 31 mm Hg

The lower P_{CO_2} than calculated indicates that she also has a respiratory alkalosis

So, the drug most likely to cause both a respiratory alkalosis and metabolic acidosis is salicylate

Salicylate intoxication classically leads to both a primary metabolic acidosis and respiratory alkalosis

A patient goes to his doctor complaining of reflux. The patient undergoes various tests that find a decrease in pH of the stomach and hypertrophy and hyperplasia of the gastric mucosa. The patient is diagnosed with a tumor that secretes a certain hormone. What hormone is it secreting?

Gastrin. The actions of gastrin are to increase H⁺ secretion by the gastric parietal cells and to stimulate the growth of the gastric mucosa

In order to isolate gastrin from the human blood stream, a researcher has designed a radioactive probe that is specific to the last five amino acids on the c-terminal of the protein. During the purification of the protein, he finds that he continues to have two different proteins. What is the other protein that the researcher is purifying?

Cholecystokinin (CCK). The structure of the five c-terminal amino acids are identical to those of gastrin

The following mixtures are ingested by a volunteer:

Mixture 1: Small peptide and amino acids

Mixture 2: Fatty acids

Mixture 3: Triglycerides

After each mixture, the patient's level of CCK is measured. Which mixture will have no affect on CCK levels?

Mixture 3, since triglycerides do not cross the intestinal cell membranes

A researcher studying somatostatin has designed an experiment in which he has injected a volunteer with this hormone and then measured the relative levels of each gastrointestinal (GI) hormone pre- and postinjection. For each of the following hormones, indicate the affect (either ↑/↓) of somatostatin on its level: gastrin, CCK, secretin, and gastric inhibitory peptide (GIP).

All ↓. Somatostatin inhibits the release of all of the GI hormones and also inhibits gastric hydrogen ion secretion

A 35-year-old woman has come to her doctor with the complaint of watery diarrhea and flushing for the past year. She says that she has over 10 watery stools per day and is concerned because it is starting to hinder her abilities at work. Upon further questioning, the woman also expresses concern over pain in the upper abdomen that radiates to the back. Blood analysis shows hyperglycemia, hypercalcemia, and hypokalemia. A hormone-secreting tumor is found to be the cause of this patient's symptoms. What hormone is it secreting?

VIP. This tumor is known as a VIPoma. VIP is a potent stimulator of cyclic adenosine monophosphate (cAMP) production in the gut, which leads to massive secretion of water and electrolyte (mainly potassium). They can be associated with multiple endocrine neoplasia (MEN) I syndrome

A researcher is studying the effects of flow of saliva and its electrolyte concentrations. For each of the electrolytes, the researcher measured its value when it is initially secreted from the acinus and when it is secreted from the salivary duct. He took measurements at high and low saliva flow rate and then compared them. What should the results look like? Use the chart below to show whether the respective electrolyte is at its highest or lowest concentration, respectively.

	Low flow	High flow
Na^+	Lowest	Highest
K^+	Highest	Lowest
Cl^-	Lowest	Highest
HCO_3^-	Unchanged	Unchanged

The salivary ducts reabsorb Na^+ and Cl^-, while it secretes K^+. So the longer it stays in the salivary duct the more Na^+ and Cl^- is reabsorbed and more K^+ is secreted into the saliva. HCO_3^- is the only one that does not follow this rule and it is continuously excreted

	Low flow	High flow
Na^+		
K^+		
Cl^-		
HCO_3^-		

The same researcher decides that he is going to study the pancreas now. But now he wants to know the relative concentration of the electrolytes compared to blood at high flow of pancreatic secretions. Indicate whether the concentration of electrolyte is higher or lower to that of the blood at high flow of pancreatic secretions

Relative concentration compared to blood

Na^+
K^+
Cl^-
HCO_3^-

	Relative concentration compared to blood
Na^+	Same
K^+	Same
Cl^-	↓
HCO_3^-	↑

At high flow, the pancreas secretes a fluid that is mostly Na^+ and HCO_3^-. At low flow, the pancreas secrets mostly Na^+ and Cl^-

A new device has been invented that measures contraction of the gallbladder. After injection of substance X, the device shows a powerful contraction of the gallbladder. What GI hormone would cause this action?

CCK. Not only does it cause the contraction of the gallbladder, but it also leads to relaxation of the sphincter of Oddi

A strange poison has been found that only inhibits the Na^+-K^+ pump in the basolateral membrane of the GI tract. What affect does this have on the absorption of glucose and galactose?

Glucose and galactose are absorbed through the Na^+-dependent cotransporter in the luminal membrane. If the Na^+-K^+ pump is inhibited, then the Na^+ gradient is destroyed, leading to inhibition of glucose and galactose absorption

Vibrio cholerae causes diarrhea by secreting a toxin that activates adenylate cyclase, which leads to increase of cAMP. Why does this lead to diarrhea?

The increased intracellular cAMP results in opening of the Cl^- channels in the luminal membrane. The flow of Cl^- out of the cell into the GI tract is accompanied by Na^+ and H_2O. This leads to a secretory diarrhea

A researcher is studying the sympathetic stimulation during a stressful situation. For the following GI tract activities, indicated the expected response in the following:

1. Digestion
2. Secretion
3. Motor activity

1. Lowers digestion
2. Lowers secretion
3. Lowers motor activity

A 5-year-old boy visiting his mom's laboratory accidentally ingested a toxin that destroys the epithelial cells of the intestinal brush border. Which of the enzymes will be affected by this ingestion?	1. Trehalase 2. α-Dextrinase 3. Peptidase 4. Isomaltase/sucrase 5. Lactase 6. Maltase 7. Proelastase
In the above scenario why will nothing happen to pepsin or trypsin?	Pepsin is found in the stomach, while trypsin is found in pancreatic secretions
A patient is told to swallow a pill containing a vitamin tagged with a radioactive substance. Twenty-four hours later, the patient's urine is negative for the vitamin. Name the above test and its purpose.	Schilling test. It is conducted to determine the etiology of vitamin B_{12} deficiency. In the given example, the lack of tagged vitamin B_{12} in urine suggests lack of intrinsic factor (IF) which can be verified with an oral dose of IF and radioactive vitamin B_{12}

A researcher is studying the sequela of fat-soluble vitamin deficiency. For each description below, identify the deficient vitamin.

1. Night blindness and dry skin	1. Vitamin A
2. Bendable bones in children	2. Vitamin D
3. Hypoprothrombinemia	3. Vitamin K
4. Increased fragility of RBCs	4. Vitamin E

A 25-year-old female presents to her primary care doctor with a 7 month history of (h/o) intermittent diarrhea with increasing fat in her stool, gum bleeding, and abdominal bloating. After extensive workup, she was diagnosed with celiac disease. Explain her symptoms.	Celiac disease is due to the toxic effects of gluten, a protein constituent of wheat flour, which alters the architecture of the small intestinal mucosa. In susceptible people, the villi are lost and the surface epitheliums are immature. This leads to loss of absorptive surface and deficiency of mucosal enzymes
A 7-year-old female immigrant from Argentina presents with progressing difficulty swallowing food. Her workup shows increased lower esophageal sphincter (LES) tone and abnormal peristalsis. What is she most likely suffering from? Why?	Achalasia secondarily to Chagas' disease
A 19-year-old female is brought by her mom to ED due to bizarre behavior and dysarthria. Workup showed increased ceruloplasmin. She was diagnosed with Wilson's disease.	Ceruloplasmin is the end product of copper that is incorporated in the liver and into apoceruloplasmin

What is ceruloplasmin and why is it elevated in Wilson's disease?	It is elevated in Wilson's disease because the transport of copper from the liver into bile is decreased leading to copper or rather excess ceruloplasmin
A 30-year-old male presents with halitosis, difficulty initiating swallow, and regurgitation of food days after ingestion. What is causing his presentation and where is it located?	Zenker's diverticulum. It is an outpouching that results from a defect in the muscular wall of the hypopharynx in a natural area of weakness known as *Killian's hiatus*, which is formed by the oblique fibers of the inferior pharyngeal constrictor muscle and the cricopharyngeal sphincter
A 32-year-old male with a 15-year h/o extensive alcohol use presents with foul-smelling stool and epigastric pain that radiates to the back and relieved by sitting upright or leaning forward. Physical examination shows pallor. His workup reveals normal lipase and elevated glucose. Why does he have elevated glucose?	Glucose intolerance occurs frequently in chronic pancreatitis, which is likely due to a decrease in pancreatic reserve or in insulin responsiveness
A man recently presented to his primary care provider with a complaint of impotence. Upon examination, the patient is found to also have galactorrhea. He is sent for an MRI, which confirms the diagnosis of a prolactinoma. What medication is used to treat him? Why?	Bromocriptine, it is a dopamine agonist. This enhances the action of dopamine on the anterior pituitary, which leads to a suppression of prolactin secretion
An elderly male has come to his PCP with a complaint of straining to urinate. After a thorough interview and examination, it is found that the patient suffers from benign prostatic hypertrophy. What kind of medications will the doctor most likely prescribe? Why?	5 α-Reductase inhibitors, they block the activation of testosterone to dihydrotestosterone in the prostate
A researcher is trying to find a new assay to detect the levels of growth hormone (GH) in the blood. What hormones have similar structures to GH that the researcher has to worry about when creating his assay?	Prolactin and human placental lactogen

A patient comes into the ER with complaints of sweating, heart palpitations, and recent weight loss. Tests show that the patient is suffering from hyperthyroidism. What medication would be used for treatment and why?	Propylthiouracil, it inhibits the peroxidase enzyme
If the above patient is found to have Graves' disease what would be the value of thyroid function tests (TFTs)?	↓ thyroid stimulating hormone (TSH), ↑ T_3, and ↑ T_4
A patient with uncontrolled atrial fibrillation is about to be started on amiodarone to control rhythm. What tests must be ordered before starting this medication?	PFTs, liver function tests (LFTs), and TFTs
A patient comes into his PCP complaining that she is gaining weight on her face. Her examination reveals a hypertensive female with a round face, and striae on her abdomen. Her labs are significant for hyperglycemia, high cortisol, and androgen levels. What hormone can be measured to differentiate Cushing's syndrome from Cushing's disease?	Adrenocorticotropic hormone (ACTH), it is high in Cushing's disease and low is Cushing's syndrome. Cushing's syndrome results from either administration of pharmacologic doses of glucocorticoids, or less likely, bilateral hyperplasia of the adrenal glands
If the above patient is found to have Cushing's disease, what medication can be used to treat her?	Ketoconazole, an inhibitor of steroid hormone synthesis
A patient is found to be hypertensive, hypokalemic, in metabolic alkalosis, and has decreased renin secretion. What disease is this patient suffering from?	Conn's syndrome (e.g., hyperaldosteronism) High aldosterone leads to 1. Increased Na^+ reabsorption, which increases the ECF → hypertension 2. Increased K^+ secretion 3. Increased H^+ secretion 4. Increased ECF and blood pressure (BP) leads to inhibition of rennin secretion
An 8-year-old boy is brought to the doctor by his parents who are concerned about the recent changes that he has undergone. It is found that he has all the signs of early puberty. What enzyme is most likely responsible for these changes?	21 β-Hydroxylase deficiency. Without this enzyme there is an excess secretion of aldosterone and sex hormones, but no production of cortisol and estradiol

Clinical Vignettes

A 17-year-old woman presents to the ER with syncope. While doing the initial assessment, it is found that the patient is hypertensive and has no axillary hair. The cause of the syncope is found to be secondary to her hypoglycemia. Additional labs on the patient show that she is hypokalemic, and has a metabolic alkalosis. What enzyme deficiency is responsible for the above findings?

17 α-Hydroxylase deficiency. The lack of axillary hair (and pubic hair) is a result from a lack of adrenal androgens. The hypoglycemia is from the decreased glucocorticoids. The metabolic alkalosis, hypokalemia, and hypertension are a result of the increased aldosterone.

While studying the pancreas, a researcher injects CCK into a patient. What hormone from the pancreas will be found at a higher level after the injection?

Glucagon

A 16-year-old girl is found by her parents in her bedroom passed out. They call 911. Emergency medical technicians arrive on the scene and find the girl to have a glucose level of 25. They begin an infusion of glucose and bring her to the hospital. By the time they reach the hospital, the patient wakes up. She refuses to speak and does not want to talk about the episode. Initial tests find the girl to have very high insulin levels, but very low protein C levels. What is the most likely etiology of the patient's hypoglycemia?

Exogenous injection of insulin. Protein C is a marker for insulin production. If this were a result from an insulinoma, the level of protein C would also be very high

A young man is brought to the ER poorly responsive. He is found to be hypotensive and tachypneic. His breath has a fruity smell. Labs show severe hyperglycemia, hyperkalemia, and metabolic acidosis. What is wrong with this patient? Explain the patient's symptoms.

The patient has classic case of diabetic ketoacidosis from uncontrolled diabetes mellitus
Hyperglycemia—insulin deficiency
Hypotension—ECF volume contraction resulting from high-filtered load of glucose exceeds the kidney's reabsorptive capacity
Metabolic acidosis—secondary from the excess production of ketoacids
Hyperkalemia—lack of insulin (insulin promotes K^+ reabsorption)

A researcher has become very interested in the female reproductive system. He has decided to study the hormones involved in ovulation. During what day of the menses would the researcher have to study in order to investigate the day of ovulation?

14 days before menses

While undergoing a removal of a neurological tumor, a young woman suffers damage to her hypothalamus leading to loss of gonadotropin-releasing hormone (GnRH) secretion. What hormone(s) would be directly affected by this?	Luteinizing hormone (LH) and follicle stimulating hormone (FSH)
A woman has been studying her ovulation cycle by measuring her basal body temperature. She finds that after ovulation her temperature begins to rise slowly. What hormone is responsible for this effect? How?	Progesterone. It has an effect on the hypothalamic thermoregulatory center
During the process of menses, the endometrium is sloughed off. What two hormones are responsible for the action?	Withdrawal of estradiol and progesterone
A 6-year-old boy is brought to his pediatrician by his worried parents because he is shorter than all his school mates. The boy's parents are of average height. What blood tests can the pediatrician order to rule out an endocrine cause of his short stature?	GH and insulin-like growth factor (IGF), as well as TSH and T_4. Both GH deficiency and hypothyroidism can result in deficits in linear growth
A pregnant woman with hypothyroidism asks her obstetrician if levothyroxine (synthetic T_4) is safe to take during pregnancy. What is her doctor's response?	It is safe and essential to continue synthetic T_4 therapy during pregnancy for all women with hypothyroidism. Thyroid hormones are important for fetal central nervous system (CNS) development
What happens to steroid hormone levels in patients with liver disease?	The liver metabolizes the majority of steroid hormones in the body. In liver disease, steroid levels increase because the rate of hepatic inactivation is diminished
An 8-year-old boy is brought to the doctor by his concerned parents, because he has recently undergone many changes resembling the signs of puberty. Describe the possible problems along the hypothalamic-pituitary-adrenal/gonadal axis that could be responsible for these changes?	Hypothalamus: excess GnRH Pituitary: excess FSH and LH Testes: excess testosterone Adrenal glands: excess steroid hormones Excess hormones, in general, are due to a hormone-secreting tumor in the associated organ

For the above boy, what adrenal enzyme problems may lead to similar symptoms?	21-Hydroxylase deficiency or 11 β-hydroxylase deficiency, which are the two most common causes of congenital adrenal hyperplasia (CAH). CAH is characterized by deficient cortisol and/or aldosterone and excess sex hormones. These enzyme deficiencies result in excessive steroid hormone precursors, which then enter the androgen pathway and lead to excess sex hormones
A patient complaining of weight change is found to have deranged glucose levels. Additionally, the patient is found to have a pancreatic tumor. Describe the weight and serum glucose changes in an insulinoma versus a glucagonoma.	Insulinoma: weight gain, hypoglycemia Glucagonoma: weight loss, hyperglycemia
A woman in her third trimester of pregnancy is worried that her baby has not moved in the past 36 hours or so. What hormone can be measured to determine fetal well-being?	Serum or urinary estriol. Estriol is made by the placenta from dehydroepiandrosterone-S (DHEA-S), which is made by fetal adrenal glands
An elderly widow complains muscle *twitches* and numbness and tingling around her mouth. On further questioning, the patient admits she has not been eating very well since her husband's death 10 months ago, and she does not go outside very much. What lab tests can reveal the etiology of her symptoms?	Serum Ca^{2+} and vitamin D levels may help establish a diagnosis of hypocalcemia. Her hypocalcemia most likely results from poor nutritional intake of calcium and is worsened by vitamin D deficiency from lack of exposure to UV light (vitamin D is important for calcium absorption from the intestines) Hypocalcemia symptoms include muscle cramps or tetany, perioral paresthesias, seizure, and osteoporosis
A 71-year-old woman with hypertension comes to her PCP for an annual checkup. She notes that she has recently been experiencing abdominal discomfort, constipation, heartburn, and joint pain. She takes hydrochlorothiazide and TUMS. How might her symptoms be explained?	Hypercalcemia secondary to thiazide and antacid (calcium carbonate) use. Hypercalcemia symptoms include calcium oxalate kidney stones, osteoporosis or pseudogout, constipation, peptic ulcer disease, depression, or altered consciousness. *Remember: stones, bones, abdominal moans, and psychic groans

For the above patient, what is her expected parathyroid hormone (PTH), calcitriol, and calcitonin levels?	PTH: decreased Calcitriol: decreased Calcitonin: increased, respectively
A 3-year-old female known with a deletion of phenylalanine at position Δ F508 presents with a 2-week h/o foul-smelling, large fatty stools, and difficulty gaining weight. What is the underling cause of her malabsorption?	Ninety percent of patients with cystic fibrosis have pancreatic insufficiency. The insufficiency results from thickened mucus in the pancreatic ducts, causing obstruction and ultimately autodigestion of the pancreas by activated enzymes

Suggested Readings

Guyton AC, Hall JE. Textbook of Medical Physiology. 10th ed. Philadelphia, Pennsylvania: Elsevier Saunders; 2000.

Ganong WF. Review of Medical Physiology. 22nd ed. Los Altos, California: Lange Medical Publications; 2005.

Mason RJ, Murray JF, Broaddus VC, Nadel JA. Murray and Nadel's Textbook of Respiratory Medicine. 4th ed. Philadelphia, Pennsylvania: Elsevier Saunders; 2005.

Kasper DL, Braunwald E, Fauci A, Hauser S, Longo D, Jameson JL. Harrison's Principles of Internal Medicine. 16th ed. New York, New York: McGraw-Hill, Medical Pub. Division; 2005.

Cunningham G, Leveno KJ, Bloom SL, Hauth JC, Gilstrap LC, Wenstrom KD. Williams Obstetrics. 22nd ed. New York, New York: McGraw-Hill, Medical Pub. Division; 2005.

Index

Page numbers followed by italic *f* refer to figures.

A

A bands, 13
Absolute refractory periods, 6, 7, 52, 53, 177
Absorption, 97
Accessory muscles, 74
Accommodation, 7
ACE. *See* Angiotensin converting enzyme
Acetazolamide, 91
Acetyl coenzyme A (COA), 8
Acetylcholine (ACh), 8, 10, 11, 21, 53, 121, 131, 133, 134, 174
Acetylcholinesterase (AChE), 10, 174
ACh. *See* Acetylcholine
Achalasia, 190
AChE. *See* Acetylcholinesterase
Acid titration, 109
Acid-base, 108–115
Acid-base disturbances, 187
Acidosis, 55, 76, 103, 104
   lactic, 114
   metabolic, 113, 114, 115
   respiratory, 113, 114, 115
Acinar cells, 132, 133
ACTH. *See* Adrenocorticotropic hormone
Actin, 14
Action potentials, 5–6, 7, 125
Active hyperemia, 66
Active tension, 16
Active transport, 136, 137
Acute respiratory distress syndrome (ARDS), 114
Acute tubular necrosis (ATN), 102
Adenosine, 77
Adenosine diphosphate (ADP), 3
Adenosine triphosphate (ATP), 3, 174
ADH. *See* Antidiuretic hormone
Adipocytes, 163, 164
Adipose tissue, 20
ADP. *See* Adenosine diphosphate
Adrenal cortex, 151
Adrenal gland, 151–155
Adrenal insufficiency, 183*f*
Adrenal medulla, 20, 151
Adrenocorticotropic hormone (ACTH), 146, 148, 152, 153, 162, 192
Adult hemoglobin (HbA), 83
Afferent fibers, 36, 89
Afterdischarge, 37
Afterload, 16, 56
   increased, 62*f*
Airway obstruction, 114
Airway resistance, 77
Albuminuria, 184
Albuterol, 21
Aldosterone, 65, 102, 104, 104*f*, 105, 137
   actions of, 153
Alkalosis, 103, 104, 114
   metabolic, 113, 115
   respiratory, 113, 115
Alpha waves, 40
Alpha-limit dextrin, 135
ALS. *See* Amyotrophic lateral sclerosis
Altitudes, 91
Alveolar gas, 78
Alveolar ventilation, 73
Alveolar-arterial gradient, 87
Alveoli, 73, 76, 77, 86
Amacrine cells, 26, 27
Amiloride, 104, 105
Amino acid hormones, 142
Amino acids, 99, 100, 135, 137, 163, 187
Amylase, 133
Amyotrophic lateral sclerosis (ALS), 114
Androgens, 82
Androstenedione, 151
Anesthetics, 114
Angioplasty, 177
Angiotensin converting enzyme (ACE), 65, 153
Angiotensin I, 65
Angiotensin II, 65, 111
ANP. *See* Atrial natriuretic peptide
ANS. *See* Autonomic nervous system
Anterior hypothalamus, 42
Anterior pituitary, 106, 146
Anterolateral system, 24

Antidiuretic hormone (ADH), 65–66, 102, 106, 107, 108, 146, 150
  secretion of, 150
Antrum, 126
Aortic arch, 64
Aortic bodies, 88
Aortic regurgitation, 63
Aortic stenosis, 63
Aortic valve, 59, 63
Aphasia, 41
Apices, 85, 175
Apneustic center, 88
Apoprotein B, 138
Arachidonic acid metabolites, 77
ARDS. *See* Acute respiratory distress syndrome
Area 4, 39
Area 6, 39
Aromatase, 159
Arterial pressure regulation, 64–66
Arteries, 45
Arterioles, 46
Astigmatism, 25
Atelectasis, 179
ATN. *See* Acute tubular necrosis
ATP. *See* Adenosine triphosphate
Atrial depolarization, 48
Atrial fibrillation, 49
Atrial flutter, 49
Atrial natriuretic peptide (ANP), 66
Atrial pressure, 47
Atrioventricular node, 48, 50, 52, 177
Atropine, 21
Auditory nerve fibers, 31
Auditory ossicles, 29–32
Autonomic nervous system (ANS), 19–22, 118, 177
  constituent parts of, 19
  effects of, 21–22
Autoregulation, 66, 67
Axons, 19–20

B
B_1-blockade, 55
Balance, 29–32
Baroreceptor reflex, 64
Baroreceptors, 64, 89
Basal ganglia, 38
Basilar membrane, 30
  apex of, 31
  base of, 31

Benzodiazepines, 41
Beta waves, 40
Bicarbonate, 137
Big gastrin, 121
Bile, 133
Bile acid micelles, 138
Bile salts, 133, 134
Bipolar cells, 26
Bitterness, 33
Bladder, 20
Blank loop, 61f
Blind spot, 27
Blood flow, 94–97
  local control, 66
  velocity, 46, 62
Blood glucose, 164
Blood urea nitrogen (BUN), 95
Blood-brain barrier, 67–68
Blood-testes barrier, 156
Body fluids, 93
Bohr effect, 83
Bony labyrinth, 30
Botulinus toxin, 12, 13
Bradycardia, 65
Bradykinin, 67
Brain stem, 12, 67
Broca's area, 42
Brodmann area 17, 28
Bromocriptine, 191
Bronchial arteries, 85
Bronchioles, 20
Bronchoconstriction, 77
Bronchodilation, 154
Brunner's gland, 134
Buffers, 108
Bulbourethral gland, 158
BUN. *See* Blood urea nitrogen
Butoxamine, 21

C
Ca^{2+}, 15, 54, 68, 106, 144, 168, 171, 174, 195
  homeostasis, 168
  metabolism, 169f
  pathway, 145f
CAH. *See* Congenital adrenal hyperplasia
Calcitonin, 169, 170, 171, 196
Calcitriol, 169, 170, 171, 196
Calcium, 121
cAMP. *See* Cyclic adenosine monophosphate
Capacitance, 47
Capillaries, 67

Carbachol, 21
Carbohydrates, 134
Carbonic anhydrase, 108
Carboxy peptidase A, 133
Carboxy peptidase B, 133
Carboxy polypeptidase, 135
Cardia, 126
Cardiac action potential, 49–50
  peak of, 50
  phases of, 50
Cardiac cycle, 59–62
Cardiac glycosides, 56
Cardiac muscle, 54–55, 163
Cardiac output (CO), 58–59
Cardiac oxygen consumption, 55
Carotid bodies, 88
Carotid massage, 64
Carotid sinus baroreceptor response, 178
Carrier-mediated transport, 2
Catecholamine pathway, 154f
Catecholamines, 56, 151, 153
CCK. *See* Cholecystokinin
Celiac disease, 190
Cell lysis, 103
Cell membranes
  components of, 1
  transport across, 1–3
Cells of Cajal, 125
Cellular anatomy, 117f
Central chemoreceptors, 88
Central sleep apnea, 89
Cephalic phase, 132
Cerebellar cortex, 38
Cerebellum, 38
Cerebral cortex, 23
  functions, 40–42
Cerebrospinal fluid (CSF), 88
Ceruloplasmin, 190
Chagas' disease, 190
CHF. *See* Congestive heart failure
Chief cells, 131
Chloride, 137
Cholecystokinin (CCK), 120, 121, 122, 123, 127, 128, 134, 187, 189
Cholesterol, 1, 67, 68, 135
Cholesterol desmolase, 152
Cholesterol esterase, 136
Cholesterol esters, 135, 136
Choline, 8
Cholinergic muscarinic antagonists, 132
Choroid plexus, 67
Chromaffin cells, 20, 151

Chronic obstructive pulmonary disease (COPD), 114, 180, 181
Chylomicrons, 138
Chyme, 128
Chymotrypsin, 133, 135
Circadian rhythms, 41
Circular muscle, 117
Circumvallate papillae, 33
Citrate, 113
CL, 99
Clearance, 94
Clonidine, 21
CO. *See* Cardiac output
CO_2, 82, 83, 84, 89, 108, 181
COA. *See* Acetyl coenzyme A
Cochlea, 30
Collecting ducts (CDs), 100, 102, 105, 107
Coloncolic reflex, 120
Compliance, 75
Conditioned reflexes, 130
Conductance, 49
Conduction, 52–54
Conduction deafness, 32
Conduction velocity, 52
Cones, 27, 28
Congenital adrenal hyperplasia (CAH), 195
Congestive heart failure (CHF), 63
Conn's syndrome, 192
Consensual reflex, 29
Contractility, 55–56
  decreased, 58f
  increased, 57f, 62f
COPD. *See* Chronic obstructive pulmonary disease
Cor pulmonale, 86
Corpus collosum, 41
Corpus luteum, 162
Corticobulbar tract, 37
Corticopapillary osmotic gradient, 107
Corticospinal tract, 37
Corticotropin-releasing hormone (CRH), 146, 147
Cortisol, 152, 153, 162
Cotransport, 1
Countercurrent exchange, 107
Counter-transport, 1, 2
Coupled transport, 137
C-peptide, 163
Creatinine, 68, 95
CRH. *See* Corticotropin-releasing hormone

CRH, ACTH, Cortisol loop, 152f
Cribriform plate, 34, 175
Cross-bridge cycle, 16
Crossed extension reflex, 176
Crypts of Lieberkuhn, 134
CSF. *See* Cerebrospinal fluid
Cupula, 175
Curare, 12, 13, 21
Cushing reaction, 65
Cushing's disease, 192
Cyanosis, 90
Cyclic adenosine monophosphate
  (cAMP), 21, 106, 142, 171, 188, 189
  pathway, 143f
Cystic fibrosis, 196
Cytochrome P450 system, 152

D

D_2 receptors, 13
Dalton's law, 78
Dantrolene, 43
Dead space, 73, 74
Deafness, 31
Decerebrate rigidity, 40
Decorticate rigidity, 40
Deep breathing, 73
Dehydration, 130
Dehydroepiandrosterone (DHEA), 151, 162, 195
Deoxyhemoglobin, 109
Depolarization, 5, 49
  of motor end-plate, 9
α-Dextrinase, 134, 135, 190
DHEA. *See* Dehydroepiandrosterone
DHT. *See* Dihydrotestosterone
Diabetic ketoacidosis, 114, 193
Diacylglycerol, 143
Diaphragm, 74
Diarrhea, 114, 182
Diastolic pressure, 47, 56
Dietary protein, 135
Diffusing capacity, 80
Diffusing capacity of carbon monoxide
  (DLCO), 178, 179
Diffusion, 1, 68, 136, 180
Diffusion coefficient, 79
Diffusion limitation, 79
Diffusion potential, 4
Diffusion trapping, 112
Digitalis, 186
Dihydrotestosterone (DHT), 155, 158
Dipeptides, 135

Direct pathway, 38, 39
Distal convoluted tubule (PCT), 100
Distal tubule, 105, 106, 107
Diuretics, 101
DLCO. *See* Diffusing capacity of carbon monoxide
Dobutamine, 21
Dopamine, 11, 39, 149
Dorsal respiratory group (DRG), 88
Dorsal root, 23
Dorsal-column system, 24
DRG. *See* Dorsal respiratory group
Dromotropic effect, 52
Duct cells, 132, 133
Duodenocolic reflex, 128
Duodenum, 122, 125, 128
Dyspnea, 90

E

ECF. *See* Extracellular fluid
Edema, 69, 184
EEG potentials. *See* Electroencephalogram potentials
Effective refractory period (ERP), 53
Efferent ductules, 156
Efferent outflow, 88
Effusion, 86
Ejection fraction, 55
Elastance, 75
Elastase, 133
Elastic work, 74
Elasticity, 75
Electrocardiogram, 47–49, 48f
  parts of, 47
Electroencephalogram (EEG)
  potentials, 40
Electrolytes, 105–106, 189
Electrophysiology, 47–49, 49–50
Embolism, 86
Emmetropia, 25
Emphysema, 178
Emulsification, 136
Endocrine hormones, 120
Endocrine pathways, 141
Endocrine system, 121, 122, 123
Endolymph, 30
Endorphins, 147
Enkephalins, 124, 147
Enterohepatic circulation, 134
Epi. *See* Epinephrine
Epididymis, 156
Epinephrine (Epi), 11, 121, 151, 154

Epithelial cells, 117
EPSP. *See* Excitatory postsynaptic potential
Equilibrium, 32
Equilibrium potential, 5
ERP. *See* Effective refractory period
ERV. *See* Expiratory reserve volume
Esophageal stage, 126
Estradiol, 159, 162, 194
Estriol, 162
Estrogen, 158, 160, 161
Excitability, 52–54
Excitation-contraction coupling, 54
Excitatory postsynaptic potential (EPSP), 10
Exercise, 74, 89–90, 103
Exocytosis, 8, 138
Expiration, 75, 88
Expiratory reserve volume (ERV), 71, 72
External intercostals muscles, 74
Extracellular fluid (ECF), 93, 101, 182
Extrafusal fibers, 35
Extrapyramidal tracts, 37
Extrinsic system, 118, 119
Eye, 20

F

Facilitated diffusion, 1, 136, 137
Fast pain, 25
Fats, 135, 138
Fat-soluble vitamins, 135, 139
Fatty acids, 163, 164, 187
Fe^2, 80
Fear, 130
Female sex organs, 158–162
FENa, 102, 103
Ferric reductase, 137, 138
Ferroxidase, 137, 138
Fetal hemoglobin (HbF), 83
Fetus, 162
FEV. *See* Forced expiratory volume
FF. *See* Filtration fraction
Fibrosing mediastinitis, 86
Fick's law, 2
Filtration fraction (FF), 96
Firing time, 35
5-HT. *See* 5-Hydroxytryptamine
5-Hydroxytryptamine (5-HT), 67
Flavoring, 32
Flexor-withdrawal reflex, 36, 37
Flow, 84
Flow limitation, 79

Fluid compartments, 182f
Follicle stimulating hormone (FSH), 146, 147, 148, 161, 162
Follicular phase, 159
Force generation, 35
Forced expiratory volume (FEV), 72, 73
Forced vital capacity (FVC), 72, 73
Formaldehyde, 113
Formate, 113
Fovea, 27
FRC. *See* Functional residual capacity
Free nerve endings, 35
Fructose, 134, 138
FSH. *See* Follicle stimulating hormone
Functional residual capacity (FRC), 71, 72
Fundus, 126
Fungiform papillae, 33
FVC. *See* Forced vital capacity

G

G cells, 121
GABA, 12
Gag, 129
Galactose, 134, 138, 189
Gallbladder, 122, 133, 189
Ganglion cells, 26
Gap junctions, 3, 54
Gas exchange/transport, 78–84
  diseases affecting, 84
Gastric distension, 129
Gastric emptying, 127
Gastric gland, 130
Gastric inhibitory peptide (GIP), 120, 123
Gastric phase, 132
Gastrin, 120, 121, 128, 131, 132, 187
Gastrin-releasing peptide (GRP), 124
Gastrocolic reflex, 120
Gastroileal reflex, 120, 128
Gastrointestinal absorption, 136–139
Gastrointestinal digestion, 134–136
Gastrointestinal hormones, 120–124
Gastrointestinal secretions, 129–134
Gastrointestinal tract, 20, 117–120
Genetic sex, 155
GFR. *See* Glomerular filtration rate
GH. *See* Growth hormone
GIP. *See* Gastric inhibitory peptide
Globus pallidus, 39
Glomerular filtration barrier, 96
Glomerular filtration rate (GFR), 95, 96, 99
Glomerulotubular balance, 101

Glomerulus, 97
Glossopharyngeal nerve, 64, 89
Glucagon, 123, 133, 163, 164, 165, 193
Glucocorticoid hormones, 77, 151
Glucose, 97, 99, 100, 134, 137, 138, 163, 164, 189, 191
Glucose titration, 98f
Glutamate, 11, 12
Glutamate receptors, 12
Glutamine, 112
Glycine, 12
Glycolipids, 1
Glycoprotein hormones, 147
GnRH. *See* Gonadotropin-releasing hormone
Golgi tendon organs, 35
Golgi tendon reflex, 36
Gonadal sex, 155, 158
Gonadotropin-releasing hormone (GnRH), 146, 147, 161
G-proteins, 142
Granular cells, 73
Granular layer, 38
Granulosa cell, 158
Gravity, 85
Growth hormone (GH), 82, 146, 147, 148, 162
    feedback loop, 149f
    secretion of, 149
GRP. *See* Gastrin-releasing peptide
Guillain-Barré syndrome, 114

H
H^+, 111, 112
    secretion, 131f
H bands, 13
$H_2PO_4^-$, 109
Hair cells, 31
Haustrations, 128
HbF. *See* Fetal hemoglobin
HCG. *See* Human chorionic gonadotropin
HCO_3^-, 68, 84, 99, 100, 189
    reabsorption of, 110f, 111
Hearing, 29–32
Heart, 20, 67
Heart failure, 55
Heart rate (HR), 55
Heart sounds, 63–64, 178
Heat exhaustion, 43
Heat stroke, 43
Heme, 80
Hemianopia, 28
Hemicholinium, 12, 13
Hemiretina, 27
Hemodynamics, 46–47
Hemoglobin, 80, 81, 82, 83, 90
Hemoglobin-oxygen dissociation curve, 81
Henderson Hasselbach equation, 109
Hereditary spherocytosis, 46
Hexamethonium, 21, 174
Hippocampus, 42
His-Purkinje system, 50
Histadine, 12
Histamine, 12, 67, 123, 124, 131, 132
Horizontal cells, 26, 27
Hormone receptor regulation, 145
Hormone secretion regulation, 145
Hormones, 141, 194
HPG axis. *See* Hypothalamus-pituitary-gonadal axis
HPL. *See* Human placental lactogen
Human chorionic gonadotropin (hCG), 147, 162
Human placental lactogen (HPL), 147, 191
Huntington's disease, 39
Hyperaldosteronism, 104, 114, 192
Hypercalcemia, 195
Hypercapnea, 55
Hyperemia, 66
Hyperglycemia, 193
Hyperkalemia, 173, 193
Hyperopia, 26
Hyperosmolarity, 103
Hyperpolarization, 5, 49
Hyperproteinemia, 46
Hyperreflexia, 40
Hypersensitivity response, 77
Hypertension, 65
Hypoaldosteronism, 104
Hypokalemia, 48
Hypoproteinemia, 184
Hyposmolarity, 103
Hypothalamic-hypophysial portal system, 147
Hypothalamus, 12, 22, 41, 146–150, 194
    anterior, 42
Hypothalamus-pituitary-gonadal (HPG) axis, 157f, 161f
Hypothermia, 43
Hypotonic, 173
Hypoventilation, 180
Hypovolemia, 76

Hypoxemia, 76, 181
Hypoxia, 46, 55, 86, 91
Hypoxic pulmonary vasoconstriction, 91

I

I band, 13
IC. *See* Inspirational capacity
ICF. *See* Intracellular fluid
Ileogastric reflex, 120, 127
Impermeability, 4
Inactivation gates, 6
Indirect pathway, 38, 39
Infant respiratory distress syndrome (IRDS), 76
Inhibitory postsynaptic potential (IPSP), 10, 11
Inner ear, 30, 30f
Inositol triphosphate (IP_3), 143
 pathway, 144f
Inotropism, 55
Inspiration, 87
 muscles in, 74
Inspirational capacity (IC), 71
Inspiratory reserve volume (IRV), 71, 72
Insulin, 103, 123, 128, 133, 163, 193
 secretion of, 164
Insulin receptors, 164
Insulin-like growth factor (IGF), 147, 194
Integral proteins, 1
Intercalated disk, 54
Interstitial fluid, 93
Intestinal brush border, 134
Intracellular connections, 3
Intracellular fluid (ICF), 93, 103, 183
Intrafusal fibers, 35
Intrapleural pressure, 75
Intrinsic factor, 139, 190
Intrinsic reflex, 129
Intrinsic system, 118, 119
Inulin, 100
Iodide, 165
Iodine, 165
Ion channels, 4–5
IP_3. *See* Inositol triphosphate
Ipsilateral telencephalon, 34
IPSP. *See* Inhibitory postsynaptic potential
IRDS. *See* Infant respiratory distress syndrome
Iron, 114, 137
Islet cells, 132, 133
Islets of Langerhans, 164
Isomaltase, 134, 190

Isometric contraction, 16
Isoniazid, 114
Isoproterenol, 21
Isosmotic sodium bicarbonate solution, 133
Isotonic contraction, 16
Isotonic saline, 182f
Isovolumetric contraction, 59, 60
Isovolumetric relaxation, 59, 60

J

Jacksonian seizure, 39
Jejunum, 122, 138
Juxtaglomerular complex, 65, 153

K

K^+, 68, 103, 105, 163, 189
Kainite receptors, 12
Ketoacids, 108, 163, 164
Ketoconazole, 192
Kidney, 20, 67, 170
Kinocilium, 32
Kussmaul breathing, 113

L

Lactase, 134, 135, 190
Lactate, 90, 100
Lactation, 162
Lactic acidosis, 114
Lactic acids, 108
Lactose, 134, 135
Lamina propria, 117
Laplace's law, 75
Large motoneurons, 35
Left ventricle, 85
Length-tension relationship, 56
Lens, 25
LES. *See* Lower esophageal sphincter
Leu-enkephalin, 124
LFTs. *See* Liver function tests
LH. *See* Luteinizing hormone
Ligand gates, 5
Ligand-gated channels, 9
Lipotropins, 147
Lithium, 150, 186
Little gastrin, 121
Liver, 133, 163, 164, 194
Liver function tests (LFTs), 192
Longitudinal muscle, 117
Long-term memories, 42
Loop diuretics, 104, 106, 114
Loop of Henle, 107

Lower esophageal sphincter (LES), 118, 190
Lumen, 117, 123, 132
Luminal anions, 104
Lungs, 65
  fetal, 76
  function of, 86
  volumes and capacities, 71–74
Luteal cell, 158
Luteal phase, 159, 160
Luteinizing hormone (LH), 146, 147, 148, 161, 162, 194
Lymph, 69
Lymphatics, 67–69

M
M line, 13
Macula densa, 153
Male sex organs, 155–158
Malignant hyperthermia, 43
Maltase, 134, 135, 190
Maltose, 135
Many-to-one synapses, 10
MAO. See Monoamine oxidase
MAP. See Mean arterial pressure
Mass movement, 128
Mast cells, 124
Mean arterial pressure (MAP), 47, 176
Mechanoreceptors, 24–25
Medulla, 22, 129
Medullary reticulospinal tract, 37
Meissner's corpuscle, 24
Melatonin, 12
Membrane potentials, 4–7
MEN I syndrome. See Multiple endocrine neoplasia I syndrome
Menstrual cycle, 159, 160f
Menstruation, 160
Merkel's disk, 24
Met-enkephalin, 124
Methanol, 113, 114
Methemoglobin, 84
Metoprolol, 21
Mg^{2+}, 68, 106
Micelles, 136, 139
Microcirculation, 67–69
Microvilli, 136
Midbrain, 22
Midline, 23
MIF. See Müllerian inhibiting factor
Mineralocorticoids, 151

MIT. See Monoiodotyrosine
Mitral cells, 34
Mitral regurgitation, 63
Mitral stenosis, 63
Mitral valve, 59, 63
Mitral valve prolapse (MVP), 63
Mobitz type 1, 48
Mobitz type 2, 48
Molecular layer, 87
Monoamine oxidase (MAO), 155
Monoiodotyrosine (MIT), 165
Monosaccharides, 134
Motility, 119, 124–129, 127
Motoneuron pool, 35
Motoneurons, 35
Motor aphasia, 41, 176
Motor homunculus, 39
Motor systems, 35–40
Motor units, 35
MS. See Multiple sclerosis
Mucous, 134
Mucous secretions, 130, 134
Müllerian inhibiting factor (MIF), 155
Multiple endocrine neoplasia (MEN) I syndrome, 188
Multiple sclerosis (MS), 114
Murmurs, 63–64
Muscarine, 21
Muscarinic receptors, 20, 21, 53, 56
Muscle
  skeletal, 13–17
  smooth, 13–17
Muscle contraction, 15–16, 35
  smooth, 17
Muscle relaxation, 16
Muscle sensors, 35
Muscle spindles, 35, 36
Muscularis mucosa, 117
MVP. See Mitral valve prolapse
Myasthenia gravis, 13
Mydriasis, 154
Myelinated fibers, 7
Myenteric plexus, 119
Myocardium, 58
Myoepithelial cells, 130
Myofibrils, 13
Myogenic mechanism, 94
Myoglobin, 83
Myopia, 26
Myosin, 14, 17

Index

N
Na^+, 68, 99, 101, 189
NaCl, 101, 107, 183
Na^+-K^+ pump, 3, 103
NaOH, 109
NE. *See* Norepinephrine
Negative feedback, 132, 146
Neostigmine, 12, 13
Nephrogenic diabetes insipidus, 186
Nephrons, 100, 107
Nephrotic syndrome, 184
Nerve cells, resting membrane potential, 6
Nerve deafness, 32
Neural crest cells, 151
Neurocrine hormones, 120
Neurocrine system, 122
Neuroendocrine hormones, 120
Neuromuscular junction, 12–13
Neuromuscular transmission, 8–13
Neurophysins, 150
Neurotransmitters
  binding of, 8
  excitatory, 11
  PNS, 20
  SNS, 20
NH_3, 112
NH_4^+, 112
Nicotine, 21
Nicotinic receptors, 8, 21
Nociception, 25
Nodes of Ranvier, 7
Nonselective beta-blockers, 179
Norepinephrine (NE), 11, 21, 151, 154
Nuclear bag fibers, 36
Nuclear chain fibers, 36
Nystagmus, 29

O
O_2, 80, 81, 82, 83, 89, 90
  capacity, 81
Obesity, 164
Obstructive sleep apnea (OSA), 89
Occupational asthma, 179
Ohm's law, 46
Olfaction, 32–40
  transduction of, 34
Olfactory epithelium, 34
Olfactory receptors, 34
Oncotic pressure, 4
1, 25-Dihydroxycholecalciferol, 137
One-to-one synapses, 10
One-way flap valves, 69
Oogenesis, 159, 160*f*
Opiates, 114
Opsin, 28
Optic chiasm, 27, 174
Optic disk, 27
Organ of Corti, 30, 31
OSA. *See* Obstructive sleep apnea
Osmolarity, 3
Osmoreceptors, 106
Osmosis, 3–4
Osmotic pressure, 4
Ovaries, 158–162
Overshoot, 6
Ovulation, 159, 162
Oxygen affinity, 82
Oxytocin, 146, 150
  secretion of, 150

P
P wave, 47, 48
Pacemaker potential, 50–52, 51*f*
  phases of, 51–52
Pacinian corpuscle, 25, 35
PAH. *See* Paraaminohippuric acid
Pain receptors, 25
Pancoast tumors, 175
Pancreas, 132, 163–165
Pancreatic amylase, 135
Pancreatic islet cell tumors, 124
Pancreatic lipase, 136
Pancreatic secretions, 135
Pancreatitis, 191
Paraaminohippuric acid (PAH), 94, 98, 99, 185
  titration of, 98
Paracrine hormones, 120
Paracrine pathways, 141
Paraldehyde, 114
Parasympathetic defecation reflex, 129
Parasympathetic nervous system (PNS), 19, 52
  neurotransmitters, 20
Parasympathetic stimulation, 55, 177
Parathyroid gland, 168–171
Parathyroid hormone (PTH), 106, 137, 169, 171
Paraventricular nuclei, 150
Parietal cells, 131
Parkinson's disease, 13, 39

Parotid gland, 129
Partial pressures, 78
Passive diffusion, 136
Passive tension, 16
Patent ductus arteriosus (PDA), 64
P_{CO_2}, 111
PCT. *See* Distal convoluted tubule; Proximal convoluted tubule
PCWP. *See* Pulmonary capillary wedge pressure
PDA. *See* Patent ductus arteriosus
Pelvic nerves, 118
Pepsin, 135, 190
Peptic cells, 131
Peptidase, 135
Peptidergic neurons, 19
Perfusion, 86–87
Perilymph, 30
Peripheral chemoreceptors, 65, 88, 89
Peripheral proteins, 1
Peripheral resistance, 56
Peristalsis, 126
Peritubular capillaries, 96
Permeability, 2
Pernicious anemia, 139
PFTs. *See* Pulmonary function tests
Pharyngeal stage, 126
Phenformin, 114
Phenotypic sex, 155, 158
Phenoxybenzamine, 21
Phentolamine, 21
Phenylalanine, 196
Phenylephrine, 21
Phosphate, 100, 109, 113
Phospholipase, 133, 136
Phospholipase C, 143
Phospholipids, 1, 135, 136
Phosphoric acid, 108
Photoreceptors, 23
Phototransduction, 28–29
Phrenic nerves, 74
Physiologic shunt, 78
PIF. *See* Prolactin inhibitory factor
Pigment epithelial cells, 26, 27
Pinocytosis, 68
Pituitary, 65, 146–150, 194
Placenta, 158–162, 162
Plasma, 93, 94
Plateau, 81
Pneumocytes, 73
Pneumonia, 114
Pneumotoxic center, 88

PNS. *See* Parasympathetic nervous system
P_{O_2}, 78, 87, 181
Poiseuille's equation, 45
Poiseuille's law, 77
Polio, 114
Polycythemia, 46
Polypeptide hormone receptors, 142
Polypeptide hormones, 141, 142, 150
POMC. *See* Proopiomelanocortin
Pons, 22, 88
Pontine reticular formation, 40
Pontine reticulospinal tract, 37
Positive feedback, 146
Posterior pituitary, 106, 146
Postganglionic nerve axons, 19–20
Postsynaptic cells, 8, 10
PPIs. *See* Proton pump inhibitors
PR interval, 47, 48
Prazosin, 21
Precapillary sphincter, 67
Preganglionic nerve axons, 19–20
Preload, 56, 61
  increased, 61*f*
Prepiriform cortex, 34
Presbyopia, 26
Pressure, 84
Pressure-volume loop, 60*f*
Presynaptic cells, 8
Primary active transport, 1
Primary lining cells, 73
Primordial follicle, 158
Principal cells, 102, 105
PRL. *See* Prolactin
Proelastase, 135, 190
Progesterone, 161, 162, 194
Prolactin (PRL), 146, 148, 162, 191
  secretion of, 149
Prolactin inhibitory factor (PIF), 146
Proopiomelanocortin (POMC), 147, 148*f*
Propranolol, 22
Proprioceptors, 89
Propylthiouracil, 192
Prostacyclin, 67
Prostaglandins, 67
Prostate, 158
Protein, 67, 68
Protein C, 193
Proton pump inhibitors (PPIs), 132
Proximal convoluted tubule (PCT), 100
Proximal jejunum, 128
Proximal tubule, 106, 110

PTH. *See* Parathyroid hormone
Ptyalin, 135
Puberty, 158
Pulmonary arteries, 85
Pulmonary blood flow, 85, 86, 177
Pulmonary capillaries, 82
Pulmonary capillary wedge pressure (PCWP), 47
Pulmonary circulation, 84–86
Pulmonary embolus, 114
Pulmonary function tests (PFTs), 72
Pulmonary gas diffusion, 79
Pulmonary vasculature, 46
Pulmonary wedge pressure, 85
Pulmonic valve, 63
Purkinje cell layer, 38
Purkinje system, 52
Pyloric gland, 130
Pylorus, 126
Pyramidal tracts, 37
Pyrogen, 42, 43

Q
QRS complex, 47, 48
QT interval, 47, 48

R
RAP. *See* Right atrial pressure
Rapid eye movement (REM), 41
RAS. *See* Renal artery stenosis
RBF. *See* Renal blood flow
Reabsorption, 97
HCO_3^-, 110f, 111
reactive hyperemia, 66
Receptive fields, 22
Receptive relaxation, 126
Receptors, 26
Red nucleus, 40
Reflection coefficient, 4
Refractory periods, 6, 53f
Relative refractory period (RRP), 7, 53
Relay nucleus, 23
REM. *See* Rapid eye movement
Renal artery stenosis (RAS), 88
Renal blood flow (RBF), 94–95, 185
Renal excretion, 97–106
Renal filtration, 94–97
Renal plasma flow (RPF), 94, 95
Renal reabsorption, 97–106
Renal secretion, 97–106
Renin, 65

Renin-angiotensin-aldosterone system, 65, 153
Renshaw cells, 37
Repolarization, 49
Reserve volume, 72
Resistance, 45, 84
Respiration, muscles of, 75
Respiratory control, 87–89
Respiratory dead space, 73
Respiratory mechanics, 74–78
Resting membrane potential, 5, 49
  for nerve cells, 6
Restrictive lung disease, 73, 178, 180
Retching, 129
Rete testis, 156
Retina, 175
Rhodopsin, 28
Right atrial pressure (RAP), 176
Rinne's test, 31
Rods, 27, 28
RPF. *See* Renal plasma flow
RRP. *See* Relative refractory period
Ruffini's corpuscle, 25

S
Saccule, 32
Salicylates, 114
Salivary ducts, 188
Salivation, 129, 130
Saltatory conduction, 7, 7f
Saltiness, 33
Sarcomeres, 13, 14f, 54
Sarcoplasmic reticulum (SR), 13, 14, 15, 54
Scala media, 30
Scala tympani, 30
Scala vestibuli, 30
Scalenes, 75
Schilling test, 190
Schizophrenia, 13
SCM. *See* Sternocleidomastoid
Scrotum, 156
Secondary sexual characteristics, 157
Second-order neurons, 23
Secretin, 120, 122, 123, 124, 133
Secretion, 97
Sedatives, 114
Segmentation, 125
Semicircular canals, 30, 32
Seminal vesicles, 158
Seminiferous tubules, 156
Sensory aphasia, 41

Sensory cortex, 24
Sensory homunculus, 24
Sensory nerve fibers, 22
Sensory receptors, 22
  tonic, 23
Sensory systems, 22–25
Sensory transduction, 23
Serosa, 117
Serotonin, 11, 12, 128
Serous secretions, 130
Sex organs, 20
SGLT-1. *See* Sodium-dependent glucose transporter
Shallow breathing, 73
Shivering, 42
Short-term memories, 42
Shunts, 87, 179
SIADH. *See* Syndrome of inappropriate antidiuretic hormone
Sigmoid curve, 81–82
Simple diffusion, 1, 137
Sinoatrial node, 50
Sinusoids, 68
Size principle, 35
Skeletal muscle, 13–17, 46, 163
Sleep, 41, 130
Slow pain, 25
Slow waves, 40, 124, 125
Small intestine, 127, 128, 129, 136
Small motoneurons, 35
Smell, 130
Smooth muscle, 13–17
  components of, 16
  contraction in, 17
  multi-unit, 17
  single-unit, 17
  vascular, 20
SNS. *See* Sympathetic nervous system
Sodium, 137
Sodium glucose/galactose cotransporter, 138
Sodium-dependent glucose transporter (SGLT-1), 138
Solvent drag, 136, 137, 138
Somatosensory system, 24
Somatostatin (SS), 123, 124, 188
Sound, transduction of, 31
Sourness, 33
Spatial summation, 9, 9f
Spermatogenesis, 156
Sphincter of Oddi, 122, 134, 189

Sphingolipids, 1
Spike potentials, 124, 125
Spinal cord, 23, 118
  transection of, 40
Spinal cord ganglia, 23
Spinal shock, 176
Spirogram, 72f
Spironolactone, 104, 105
Splay, 98
SR. *See* Sarcoplasmic reticulum
ST segment, 47, 48
Standing curves, 56
Starch, 134, 135
Starling curve, 57f
Starling equation, 68, 95, 96
Starling forces, 95
Starling relationships, 56–58
Starvation state, 164
Steep section, 81
Sternocleidomastoid (SCM), 75
Steroid hormone receptors, 142
Steroid hormones, 141, 145, 153
Stomach
  anatomy of, 126, 127f
  functions of, 127
strap muscles of neck, 75
Stress, 89–91
Stretch receptors, 64
Stretch reflex, 36, 37
Striatum, 39
Stroke volume (SV), 47
Sublingual gland, 129
Submandibular gland, 129
Submucosa, 117
Submucosal plexus, 119
Substance P, 25, 77
Substantia nigra, 39
Subthalamic nuclei, 39
Sucrase, 135, 190
Sucrose, 134, 135
Sulfate, 113
Supplementary motor cortex, 39
Suprachiasmatic nucleus, 41
Surfactant, 76, 179
SV. *See* Stroke volume
Swallow center, 126
Swallow reflex, 126
Swan-Ganz catheter, 47
Sweat glands, 20, 42
Sweating, 184, 185f
Sweetness, 33

Sympathetic nervous system (SNS), 19, 118
　neurotransmitters, 20
　receptor types, 20–21
Synaptic cleft, 8
Synaptic transmission, 8–13
Synaptic vesicles, 8
Syndrome of inappropriate antidiuretic hormone (SIADH), 107, 184, 185
Synthetic pathways, 11
Synthetic T, 194
Systemic circulation, 84
Systemic vasculature, 46
Systolic pressure, 47, 85

T

T tubule, 54
T wave, 47, 48
T_3, 165, 166, 167
T_4, 165, 166, 167
Tachycardia, 90
Tachypnea, 90
Taste, 32–40
　pathway for, 34
Taste buds, 33
TBW. *See* Total body water
TDF. *See* Testicular differentiation factor
Tectorial membrane, 30
Temperature regulation, 42–43
Temporal summation, 9, 10*f*
Tension maximum, 16
Terminal ileum, 128, 139
Testicular differentiation factor (TDF), 155
Testis, 155–158, 194
Testosterone, 156, 158
　actions of, 157
Thalamus, 23, 38, 174
Theca cell, 158
Thiazide diuretics, 101, 104, 106, 114
Thick filaments, 13, 14, 16
Thin filaments, 13, 14, 16
Threshold, 10
Thyroglobulin, 165
Thyroid gland, 146, 165–168
Thyroid hormone, 82, 145, 165
　secretion of, 166
　synthesis, 166*f*
Thyroid stimulating hormone (TSH), 146, 148, 167
Thyrotropin-releasing hormone (TRH), 146, 167
Tidal volume, 71

Tight junctions, 3
Tissue capillaries, 82
Tissue protein, 135
Titration, 98
TLC. *See* Total lung capacity
Total body water (TBW), 93
Total lung capacity (TLC), 71, 72
Total tension, 16
Transport
　carrier-mediated, 2
　types of, 1
Transport maximum, 98
Transverse tubules, 13, 15
Trehalase, 135, 190
TRH. *See* Thyrotropin-releasing hormone
TRH, TSH, and thyroid hormone loop, 168*f*
Triamterene, 104, 105
Tricuspid valve, 63
Triglycerides, 135, 136, 187
Tripeptides, 135
Tropomyosin, 14
Troponin, 14
Trypsin, 133, 135
Tryptophan, 12
TSH. *See* Thyroid stimulating hormone
Tubuloglomerular feedback, 94
Tympanic membrane, 29
Type II alveolar epithelial cells, 76
Tyrosine, 153

U

UES. *See* Upper esophageal sphincter
Undershoot, 6
Upper esophageal sphincter (UES), 129
Urea, 106
Urea recycling, 107
Uremia, 114
Urethral gland, 158
Urine, 97, 106–108
Utricle, 32

V

Vagovagal reflex, 119
Vagus nerve, 64, 89, 118, 124, 132
Valvulae conniventes, 136
Vanillylmandelic acid (VMA), 155
Vas deferens, 156
Vasa recta, 107
Vascular smooth muscle, 20
Vasculature, 45–46
Vasculitis, 86

Vasoactive inhibitory peptide (VIP), 123, 124, 188
Vasoconstriction, 86
Vasodilators, 56
Vasomotor center, 64, 65
Vasopressin, 65
VC. See Vital capacity
Veins, 45
Velocity, blood flow, 46, 62
Venodilators, 56
Ventilation, 86–87
Ventral respiratory group (VRG), 89
Ventricular depolarization, 48
Ventricular ejection, 59, 60, 61
Ventricular fibrillation, 49
Ventricular filling, 59, 60
Ventricular repolarization, 48
Ventricular septal defect (VSD), 63
Vestibular stimulation, 129
Vestibule, 30
Vestibulospinal tract, 37
Villi, 136
VIP. See Vasoactive inhibitory peptide
Visceral pain, 25
Visceral smooth muscle (VSM), 118
Viscosity, 46
Viscous resistance, 74
Vision, 25–29
  neural pathway of, 27
Vital capacity (VC), 71, 72
Vitamin A, 28, 139, 175, 190

Vitamin B_{12}, 139
Vitamin D, 139, 170, 190, 195
Vitamin E, 139, 190
Vitamin K, 139, 190
VMA. See Vanillylmandelic acid
Voltage gates, 5
Voluntary stage, 126
Vomiting, 114, 129
V/Q ratio, 86, 87
VRG. See Ventral respiratory group
VSD. See Ventricular septal defect
VSM. See Visceral smooth muscle

W

Water deprivation, 107
Water-soluble vitamins, 139
Weber's test, 32
Wernicke's area, 41
Wilson's disease, 191

Y

Yohimbine, 22

Z

Z lines, 13
Zenker's diverticulum, 191
Zollinger-Ellison syndrome, 122
Zona fasciculata, 151
Zona glomerulosa, 151
Zona reticularis, 151
Zymogens, 133